# STUDENT'S SOLUTIONS MANUAL

### WILLIAM CRAINE III
*Lansing High School*

# INTRO STATS
## FIFTH EDITION

# Richard De Veaux
*Williams College*

# Paul Velleman
*Cornell University*

# David Bock
*Cornell University*

The author and publisher of this book have used their best efforts in preparing this book. These efforts include the development, research, and testing of the theories and programs to determine their effectiveness. The author and publisher make no warranty of any kind, expressed or implied, with regard to these programs or the documentation contained in this book. The author and publisher shall not be liable in any event for incidental or consequential damages in connection with, or arising out of, the furnishing, performance, or use of these programs.

Reproduced by Pearson from electronic files supplied by the author.

 Pearson

ISBN-13: 978-0-13-426535-3
ISBN-10: 0-13-426535-1

# Contents

# Chapter 1 – Stats Starts Here

## Section 1.1

1. **Grocery shopping.** Discount cards at grocery stores allow the stores to collect information about the products that the customer purchases, what other products are purchased at the same time, whether or not the customer uses coupons, and the date and time that the products are purchased. This information can be linked to demographic information about the customer that was volunteered when applying for the card, such as the customer's name, address, sex, age, income level, and other variables. The grocery store chain will use that information to better market their products. This includes everything from printing out coupons at the checkout that are targeted to specific customers to deciding what television, print, or Internet advertisements to use.

3. **Parking lots.** The owners of the parking garage can advertise about the availability of parking. They can also communicate with businesses about hours when more spots are available and when they should encourage more business.

## Section 1.2

5. **Super Bowl.** When collecting data about the Super Bowl, the games themselves are the *Who*.

7. **Health records.** The sample is about 5,000 people, and the population is all residents of the United States of America. The *Who* is the selected subjects and the *What* includes medical, dental, and physiological measurements and laboratory test results.

## Section 1.3

9. **Grade level.**

   a) If we are, for example, comparing the percentage of first-graders who can tie their own shoes to the percentage of second-graders who can tie their own shoes, grade-level is treated as categorical. It is just a way to group the students. We would use the same methods if we were comparing boys to girls or brown-eyed kids to blue-eyed kids.

   b) If we were studying the relationship between grade-level and height, we would be treating grade level as quantitative.

11. **Voters.** The response is a categorical variable.

13. **Medicine.** The company is studying a quantitative variable.

**Section 1.4**

**15. Voting and elections.** Pollsters might consider whether a person voted previously or whether he or she could name the candidates. Voting previously and knowing the candidates may indicate a greater interest in the election.

**17. The News.** Answers will vary.

**19. Gaydar.** *Who* – 40 undergraduate women. *What* – Whether or not the women could identify the sexual orientation of men based on a picture. *Population of interest* – All women.

**21. Bicycle Safety.** *Who* – 2,500 cars. *What* – Distance from the bicycle to the passing car (in inches). *Population of interest* – All cars passing bicyclists.

**23. Honesty.** *Who* – Workers who buy coffee in an office. *What* – amount of money contributed to the collection tray. *Population of interest* – All people in honor system payment situations.

**25. Not-so-diet soda.** *Who* – 474 participants. *What* – whether or not the participant drank two or more diet sodas per day, waist size at the beginning of the study, and waist size at the end of the study. *Population of interest* – All people.

**27. Weighing bears.** *Who* – 54 bears. *What* – Weight, neck size, length (no specified units), and sex. *When* – Not specified. *Where* – Not specified. *Why* - Since bears are difficult to weigh, the researchers hope to use the relationships between weight, neck size, length, and sex of bears to estimate the weight of bears, given the other, more observable features of the bear.
*How* – Researchers collected data on 54 bears they were able to catch. *Variables* – There are 4 variables; weight, neck size, and length are quantitative variables, and sex is a categorical variable. No units are specified for the quantitative variables. *Concerns* – The researchers are (obviously!) only able to collect data from bears they were able to catch. This method is a good one, as long as the researchers believe the bears caught are representative of all bears, in regard to the relationships between weight, neck size, length, and sex.

**29. Arby's menu.** *Who* – Arby's sandwiches. *What* – type of meat, number of calories (in calories), and serving size (in ounces). *When* – Not specified. *Where* – Arby's restaurants. *Why* – These data might be used to assess the nutritional value of the different sandwiches. *How* – Information was gathered from each of the sandwiches on the menu at Arby's, resulting in a census. *Variables* – There are three variables. Number of calories and serving size (ounces) are quantitative variables, and type of meat is a categorical variable.

31. **Babies.** *Who* – 882 births. *What* – Mother's age (in years), length of pregnancy (in weeks), type of birth (caesarean, induced, or natural), level of prenatal care (none, minimal, or adequate), birth weight of baby (unit of measurement not specified, but probably pounds and ounces), gender of baby (male or female), and baby's health problems (none, minor, major). *When* – 1998-2000. *Where* – Large city hospital. *Why* – Researchers were investigating the impact of prenatal care on newborn health. *How* – It appears that they kept track of all births in the form of hospital records, although it is not specifically stated. *Variables* – There are three quantitative variables: mother's age (years), length of pregnancy (, and birth weight of baby. There are four categorical variables: type of birth, level of prenatal care, gender of baby, and baby's health problems.

33. **Herbal medicine.** *Who* – experiment volunteers. *What* – herbal cold remedy or sugar solution, and cold severity (0 to 5 scale). *When* – Not specified. *Where* – Major pharmaceutical firm. *Why* – Scientists were testing the efficacy of an herbal compound on the severity of the common cold. *How* – The scientists set up a controlled experiment. *Variables* – There are two variables. Type of treatment (herbal or sugar solution) is categorical, and severity rating is quantitative. *Concerns* – The severity of a cold seems subjective and difficult to quantify. Also, the scientists may feel pressure to report negative findings about the herbal product.

35. **Streams.** *Who* – Streams. *What* – Name of stream, substrate of the stream (limestone, shale, or mixed), acidity of the water (measured in pH), temperature (in degrees Celsius), and BCI (unknown units). *When* – Not specified. *Where* – Upstate New York. *Why* – Research is conducted for an Ecology class. *How* – Not specified. *Variables* – There are five variables. Name and substrate of the stream are categorical variables, and acidity, temperature, and BCI are quantitative variables.

37. **Refrigerators.** *Who* – 353 refrigerator models. *What* – Brand, cost (probably in dollars), size (in cu. ft.), type, estimated annual energy cost (probably in dollars), overall rating, and repair history (in percent requiring repair over the past five years). *When* – 2013. *Where* – United States. *Why* – The information was compiled to provide information to the readers of *Consumer Reports*. *How* – Not specified. *Variables* – There are 7 variables. Brand, type, and overall rating are categorical variables. Cost, size, estimated energy cost, and repair history are quantitative variables.

39. **Kentucky Derby 2016.** *Who* – Kentucky Derby races. *What* – Year, winner, jockey, trainer, owner, and time (in minutes, seconds, and hundredths of a second). *When* – 1875 – 2016. *Where* – Churchill Downs, Louisville, Kentucky. *Why* – Not specified. To examine the trends in the Kentucky Derby? *How* – Official statistics are kept for the race each year. *Variables* – There are 6 variables. Winner, jockey, trainer and owner are categorical variables. Date and duration are quantitative variables.

**41. Kentucky Derby 2016 on the computer.**

   **a)** Fonso was the winning horse in 1880.

   **b)** The length of the race changed in 1895, from 1.5 miles to 1.25 miles.

   **c)** The winning time in 1974 was 124 seconds.

   **d)** Secretariat ran the Derby in under 2 minutes in 1973, as did Monarchos in 2001.

# Chapter 2 – Displaying and Describing Data

## Section 2.1

1. **Automobile fatalities.**

| Subcompact and Mini | 0.2658 |
|---------------------|--------|
| Compact | 0.2084 |
| Intermediate | 0.3006 |
| Full | 0.2069 |
| Unknown | 0.0183 |

2. **Non-occupant fatalities.**

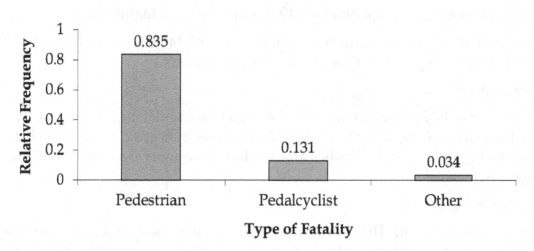

3. **Movie genres.**

   a) A pie chart seems appropriate from the movie genre data. Each movie has only one genre, and the list of all movies constitute a "whole".

   b) "Other" is the least common genre. It has the smallest region in the chart.

4. **Movie ratings.**

   a) A pie chart seems appropriate for the movie rating data. Each movie has only one rating, and the list of all movies constitute a "whole". The percentages of each rating are different enough that the pie chart is easy to read.

   b) The most common rating is R. It has the largest region on the chart.

5. **Movie ratings.**

   i) C     ii) A     iii) D     iv) B

6. **Marriage in decline.**

   i) D     ii) A     iii) C     iv) B

**Section 2.2**

7. **Traffic Fatalities 2013.**

    a) The gaps in the histogram for *Year* indicate that we do not have data for those years. This data set contains two variables for each case, and a histogram of the years doesn't give us much useful information.

    b) All of the bars in the *Year* histogram are the same height because each year only appears once in the data set.

    c) The distribution of passenger car fatalities has between 17,500 and 25,000 traffic fatalities per year in most years. There were also several years—possibly a second mode—with between 10,000 and 12,500 traffic fatalities.

8. **Traffic Fatalities 2013 again.**

    a) Two years were included in the highlighted bar in the passenger car fatalities histogram.

    b) The number of passenger car fatalities for 1975 and 1980 were highlighted.

    c) The two years with the highest number of passenger car fatalities were 1975 and 1980. There have been fewer passenger car fatalities in recent years.

9. **How big is your bicep?**

    The distribution of the bicep measurements of 250 men is unimodal and symmetric. Based on the height of the tallest points, about 85 of these 250 men have biceps close to 13 inches around. Most are between 12 and 15 inches around. But there are two as small as 10 inches and several that are 16 inches.

10. **How big is your bicep in cm?**

    Yes, the dotplots look different. The distribution of the bicep measurements in centimeters is still unimodal, but not as symmetric. The plot based on inches has fewer values on the *x*-axis, so it shows less detail. The measurements were rounded to the nearest half-inch after converting to inches, giving the plot a granular appearance. The plot in centimeters gives a better picture of the distribution.

11. **E-mails.**

    The distribution of the number of emails received from each student by a professor in a large introductory statistics class during an entire term is skewed to the right, with the number of emails ranging from 1 to 21 emails. The distribution is centered at about 2 emails, with many students only sending 1 email. There is one outlier in the distribution, a student who sent 21 emails. The next highest number of emails sent was only 8.

12. **Adoptions.**

    a) The distributions of the number of adoptions and state populations are both skewed to the right. Most states have smaller populations and fewer adoptions, but some big states have substantially more of each.

b) The distributions have similar shapes, since states with higher populations are likely to have more adoptions.

c) The number of adoptions could be expressed as a rate. For example, report the number of adoptions per 100,000 people.

**Section 2.3**

**13. Biceps revisited.**

The distribution of the bicep measurements of 250 men is unimodal and roughly symmetric.

**14. E-mails II.**

The distribution of the number of emails received from each student by a professor in a large introductory statistics class during an entire term is skewed to the right.

**15. Life expectancy.**

a) The distribution of life expectancies at birth in 190 countries is skewed to the left.

b) The distribution of life expectancies at birth in 190 countries has one mode, at about 74 to 76 years. The fluctuations from bar to bar don't seem to rise to the level of defining additional modes, although opinions can differ.

**16. Shoe sizes.**

a) The distribution of European shoe sizes of 269 college students in roughly symmetric and possibly bimodal.

b) The distribution of European shoe sizes of 269 college students seems to have two modes, one between sizes 38 and 40, and another between sizes 44 and 46. This could be due to having data for both men and women. The lower mode may be the typical shoe size for women, and the upper mode may be the typical shoe size for men.

**17. Life expectancy II.**

a) The distribution of life expectancies at birth in 190 countries is skewed to the left, so the median is expected to be larger than the mean. The mean life expectancy is pulled down toward the tail of the distribution.

b) Since the distribution of life expectancies at birth in 190 countries is skewed to the left, the median is the better choice for reporting the center of the distribution. The median is more resistant to the skewed shape of the distribution.

**18. Adoptions II.**

a) The distribution of the number of adoptions is skewed to the right, so the mean is expected to be larger. The mean number of adoptions is pulled up toward the tail of the distribution.

b) Since the distribution of the number of adoptions is skewed to the right, the median is the better choice for reporting the center of the distribution. The median is more resistant to the skewed shape of the distribution.

**19. How big is your bicep II?**

Because the distribution of bicep circumferences is unimodal and symmetric, the mean and the median should be very similar. The usual choice is to report the mean or to report both.

**20. Shoe sizes II.**

Because the distribution of shoe sizes has two modes, the mean and median are not helpful in reporting the story that the data tell. It is better to report the locations of the two modes.

**Section 2.5**

**21. Life expectancy III.**

a) We should report the IQR.

b) Since the distribution of life expectancies at birth in 190 countries is skewed to the left, the better measure of spread is the IQR. The skewness of the distribution inflates the standard deviation.

**22. Adoptions III.**

a) We should report the IQR.

b) Since the distribution of the number of adoptions is skewed to the right, the IQR is the better choice for reporting the spread of the distribution. The skewness of the distribution inflates the standard deviation.

**23. How big is your bicep III?**

Because the distribution of bicep circumferences is unimodal and roughly symmetric, we should report the standard deviation. The standard deviation is generally more useful whenever it is appropriate. However, it would not be strictly wrong to use the IQR. We just prefer the standard deviation.

**24. Shoe sizes III.**

The data combine shoe sizes for men and for women. It isn't appropriate to summarize a bimodal distribution as if they were a single collection of values.

**Chapter Exercises**

**25. Graphs in the news.** Answers will vary.

**26. Graphs in the news II.** Answers will vary.

**27. Tables in the news.** Answers will vary.

**28. Tables in the news II.** Answers will vary.

**29. Histogram.** Answers will vary.

**30. Not a histogram.** Answers will vary.

**31. Centers in the news.** Answers will vary.

**32. Spreads in the news .** Answers will vary.

## 33. Thinking about shape.

a) The distribution of the number of speeding tickets each student in the senior class of a college has ever had is likely to be unimodal and skewed to the right. Most students will have very few speeding tickets (maybe 0 or 1), but a small percentage of students will likely have comparatively many (3 or more?) tickets.

b) The distribution of player's scores at the U.S. Open Golf Tournament would most likely be unimodal and slightly skewed to the right. The best golf players in the game will likely have around the same average score, but some golfers might be off their game and score 15 strokes above the mean. (Remember that high scores are undesirable in the game of golf!)

c) The weights of female babies in a particular hospital over the course of a year will likely have a distribution that is unimodal and symmetric. Most newborns have about the same weight, with some babies weighing more and less than this average. There may be slight skew to the left, since there seems to be a greater likelihood of premature birth (and low birth weight) than post-term birth (and high birth weight).

d) The distribution of the length of the average hair on the heads of students in a large class would likely be bimodal and skewed to the right. The average hair length of the males would be at one mode, and the average hair length of the females would be at the other mode, since women typically have longer hair than men. The distribution would be skewed to the right, since it is not possible to have hair length less than zero, but it is possible to have a variety of lengths of longer hair.

## 34. More shapes.

a) The distribution of the ages of people at a Little League game would likely be bimodal and skewed to the right. The average age of the players would be at one mode and the average age of the spectators (probably mostly parents) would be at the other mode. The distribution would be skewed to the right, since it is possible to have a greater variety of ages among the older people, while there is a natural left endpoint to the distribution at zero years of age.

b) The distribution of the number of siblings of people in your class is likely to be unimodal and skewed to the right. Most people would have 0, 1, or 2 siblings, with some people having more siblings.

c) The distribution of pulse rate of college-age males would likely be unimodal and symmetric. Most males' pulse rates would be around the average pulse rate for college-age males, with some males having lower and higher pulse rates.

d) The distribution of the number of times each face of a die shows in 100 tosses would likely be uniform, with around 16 or 17 occurrences of each face (assuming the die had six sides).

## 35. Movie genres again.

a) Thriller/Suspense has a higher bar than Adventure, so it is the more common genre.

b) It is easy to tell from either chart; sometimes differences are easier to see on the bar chart because slices of the pie chart look too similar in size.

**36. Movie ratings, again.**

a) The least common rating was NC-17. It has the shortest bar.

b) It is easy to tell from either chart; sometimes differences are easier to see on the bar chart because slices of the pie chart look too similar in size.

**37. Magnet Schools.**

There were 1755 qualified applicants for the Houston Independent School District's magnet schools program. 53% were accepted, 17% were wait-listed, and the other 30% were turned away for lack of space.

**38. Magnet schools again.**

There were 1755 qualified applicants for the Houston Independent School District's magnet schools program. 29.5% were Black or Hispanic, 16.6% were Asian, and 53.9% were white.

**39. Causes of death 2014.**

a) Yes, it is reasonable to assume that heart or lung diseases caused approximately 29% of U.S. deaths in 2014, since there is no possibility for overlap. Each person could only have one cause of death.

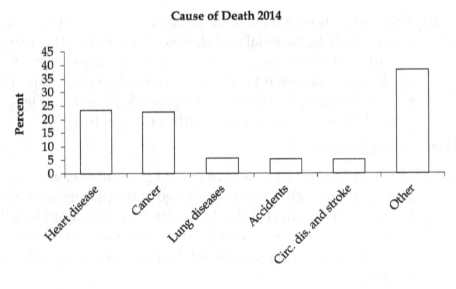

b) Since the percentages listed add up to 61.9%, other causes must account for 38.1% of US deaths.

c) A bar chart is a good choice (with the inclusion of the "Other" category). Since causes of US deaths represent parts of a whole, a pie chart would also be a good display.

**40. Plane crashes.**

a) As long as each plane crash had only one cause, it would be reasonable to assume that weather or mechanical failures were the causes of about 37% of crashes.

b) It is likely that the numbers in the table add up to 101% due to rounding.

c) A relative frequency bar chart is a good choice. A pie chart would also be a good display, as long as each plane crash has only one cause.

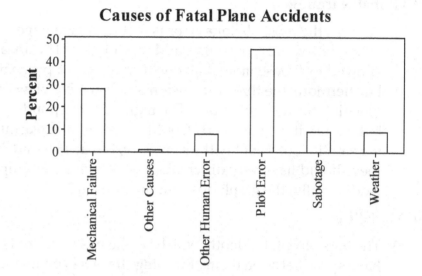

**Causes of Fatal Plane Accidents**

## 41. Movie genres once more.

a) There are too many categories to construct an appropriate display. In a bar chart, there are too many bars. In a pie chart, there are too many slices. In each case, we run into difficulty trying to display genres that only represented a few movies.

b) The creators of the bar chart included a category called "Other" for many of the genres that only occurred a few times.

## 42. Summer Olympics 2016.

a) There are too many categories to construct an appropriate display. In a bar chart, there are too many bars. In a pie chart, there are too many slices. In each case, we run into difficulty trying to display those countries that didn't win many medals.

b) Perhaps we are primarily interested in countries that won many medals. We might choose to combine all countries that won fewer than 6 medals into a single category. This will make our chart easier to read. But, we are probably more interested in the number of medals won overall, and don't need to know what countries won the medals. A histogram is probably the most appropriate display. Now we can see that most countries won very few medals, and a handful of countries won many medals per country.

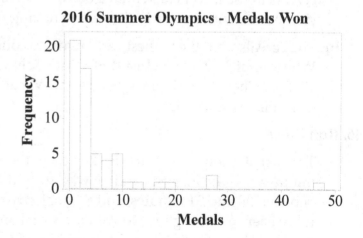

**2016 Summer Olympics - Medals Won**

### 43. Global warming.

Perhaps the most obvious error is that the percentages in the pie chart add up to 141%, when they should, of course, add up to 100%. This means that survey respondents were allowed to choose more than one response, so a pie chart is not an appropriate display. Furthermore, the three-dimensional perspective view distorts the regions in the graph, violating the area principle. The regions corresponding to "Could reduce global warming but unsure if we will" and "Could reduce global warming but people aren't willing to so we won't" look roughly the same size, but at 46% and 30% of respondents, respectively, they should have very different sizes. Always use simple, two-dimensional graphs. Additionally, the graph does not include a title.

### 44. Modalities.

a) The bars have false depth, which can be misleading. This is a bar chart, so the bars should have space between them. Running the labels on the bars from top to bottom and the vertical axis labels from bottom to top is confusing.

b) The percentages sum to 100%. Normally, we would take this as a sign that all of the observations had been correctly accounted for. But in this case, it is extremely unlikely. Each of the respondents was asked to list *three* modalities. For example, it would be possible for 80% of respondents to say they use ice to treat an injury, and 75% to use electric stimulation. The fact that the percentages total greater than 100% is not odd. In fact, in this case, it seems wrong that the percentages add up to 100%, rather than correct.

### 45. Cereals.

a) The distribution of the carbohydrate content of breakfast cereals is bimodal, with a cluster of cereals with carbohydrate content around 13 grams of carbs and another cluster of cereals around 22 grams of carbs. The lower cluster shows a bit of skew to the left. Most cereals in the lower cluster have between 10 and 20 grams of carbs. The upper cluster is symmetric, with cereals in the cluster having between 20 and 24 grams of carbs.

b) The cereals with the highest carbohydrate content are Corn Chex, Corn Flakes, Cream of Wheat (Quick), Crispix, Just Right Fruit & Nut, Kix, Nutri-Grain Almond-Raisin, Product 19, Rice Chex, Rice Krispies, Shredded Wheat 'n' Bran, Shredded Wheat Spoon Size, Total Corn Flakes, and Triples.

### 46. Run times.

The distribution of runtimes is skewed to the right. The shortest runtime was around 28.5 minutes and the longest runtime was around 35.5 minutes. A typical run time was between 30 and 31 minutes, and the majority of runtimes were between 29 and 32 minutes. It is easier to run slightly slower than usual and end up with a longer runtime than it is to run slightly faster than usual and end up with a shorter runtime. This could account for the skew to the right seen in the distribution.

## 47. Heart attack stays.

a) The distribution of length of stays is skewed to the right, so the mean is larger than the median.

b) The distribution of the length of hospital stays of female heart attack patients is bimodal and skewed to the right, with stays ranging from 1 day to 36 days. The distribution is centered around 8 days, with the majority of the hospital stays lasting between 1 and 15 days. There are a relatively few hospital stays longer than 27 days. Many patients have a stay of only one day, possibly because the patient died.

c) The median and IQR would be used to summarize the distribution of hospital stays, since the distribution is strongly skewed.

## 48. Bird species 2013.

Number of Birds

a) The results of the 2013 Laboratory of Ornithology Christmas Bird Count are displayed in the stem and leaf display at the right.

b) The distribution of the number of birds spotted by participants in the 2013 Laboratory of Ornithology Christmas Bird Count is skewed right, with a median of 117 birds. There are three high potential outliers, with participants spotting 150, 166, and 184 birds. With the exception of these outliers, most participants saw between 82 and 136 birds.

```
 8 | 2368
 9 | 78
10 | 1156
11 | 8
12 | 468
13 | 136
14 |
15 | 0
16 | 6
17 |
18 | 4
```

Key : 15 | 0 = 150 birds

## 49. Super Bowl points 2016.

a) The median number of points scored in the first 50 Super Bowl games is 46 points.

b) The first quartile of the number of points scored in the first 50 Super Bowl games is 37 points. The third quartile is 55 points.

c) In the first 50 Super Bowl games, the lowest number of points scored was 21, and the highest number of points scored was 75. The median number of points scored was 46, and the middle 50% of Super Bowls has between 37 and 55 points scored, making the IQR 18 points.

## 50. Super Bowl edge 2016.

a) The median winning margin in the first 50 Super Bowl games is 12.5 points.

b) The first quartile of the winning margin in the first 50 Super Bowl games is 4 points. The third quartile is 19 points.

c) In the first 50 Super Bowl games the lowest winning margin was 1 point and the highest winning margin was 45 points. The median winning margin was 12.5 points, with the middle 50% of winning margins between 4 and 19 points, making the IQR 15 points.

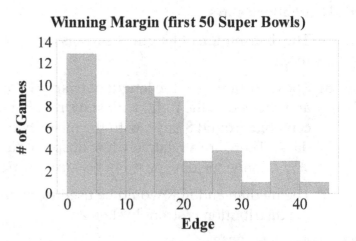

**Winning Margin (first 50 Super Bowls)**

**51. Test scores, large class.**

a) The distribution of Calculus test scores is bimodal with one mode at about 62 and one at about 78. The higher mode might be math majors, and the lower mode might be non-math majors.

b) Because the distribution of Calculus test scores is bimodal, neither the mean nor the median tells much about a typical score. We should attempt to learn if another variable (such as whether or not the student is a math major) can account for the bimodal character of the distribution.

**52. Test scores, small class.**

The distribution of Calculus test scores appears bimodal, with one mode at about 62 and one at about 78. The higher mode might be math majors, and the lower mode might be non-math majors. Because there are so few values in the middle bins, it is not as clear that this distribution is actually bimodal. The distribution might merely be skewed to the left, if one considers that it would only take several scores in the middle bars to fill in the "gap".

**53. Mistake.**

a) As long as the boss's true salary of $200,000 is still above the median, the median will be correct. The mean will be too large, since the total of all the salaries will decrease by $2,000,000 - $200,000 = $1,800,000, once the mistake is corrected.

b) The range will likely be too large. The boss's salary is probably the maximum, and a lower maximum would lead to a smaller range. The IQR will likely be unaffected, since the new maximum has no effect on the quartiles. The standard deviation will be too large, because the $2,000,000 salary will have a large squared deviation from the mean.

**54. Sick days.**

a) The company probably uses the mean, while the union uses the median number of sick days. The mean will likely be higher, since it is affected by probable right skew. Some employees may have many sick days, while most have relatively few.

**b)** These choices for summaries make sense: The company cares about the number of days lost to sickness, so the mean (related to the total) is what matters. The union looks at individual workers and sees that the typical (median) worker doesn't miss many days for sickness. Additionally, each side is trying to make their side's position in the negotiation look better. The union benefits from a perception of lower numbers of sick days, while the company benefits from the perception of a higher number of sick days, which need to be curtailed.

## 55. Floods 2015.

**a)** The mean annual number of deaths from floods is 81.95.

**b)** In order to find the median and the quartiles, the list must be ordered.
29  38  38  43  48  49  56  68  76  80  82  82  82  86 87  103  113  118  131  136  176
The median annual number of deaths from floods is 82.
Quartile 1 = 49 deaths, and Quartile 3 = 103 deaths.
(Some statisticians consider the median to be separate from both the lower and upper halves of the ordered list when the list contains an odd number of elements.  This changes the position of the quartiles slightly.   If median is excluded, Q1 = 48.5, Q3 = 108.  In practice, it rarely matters, since these measures of position are best for large data sets.)

**c)** The range of the distribution of deaths is Max − Min = 176 − 29 = 147 deaths.
The IQR = Q3 − Q1 = 103 − 49 = 54 deaths. (Or, the IQR = 108 − 48.5 = 59.5 deaths, if the median is excluded from both halves of the ordered list.)

## 56. Tornadoes 2015.

**a)** The mean annual number of deaths from tornadoes is 81.43.

**b)** In order to find the median and the quartiles, the list must be ordered.
21  25  30  35  36  38  40  41  45  47  54  55  55  67  67  70  81  94  126  130  553
The median annual number of deaths from tornadoes is 54.
Quartile 1 = 38 deaths, and Quartile 3 = 70 deaths.
(Some statisticians consider the median to be separate from both the lower and upper halves of the ordered list.  This changes the position of the quartiles slightly.   If median is included, Q1 = 37, Q3 = 75.5.  In practice, it rarely matters, since these measures of position are best for large data sets.)

**c)** The range of the distribution of deaths is Max − Min = 553 − 21 = 532 deaths.
The IQR = Q3 − Q1 = 70 − 38 = 32 deaths. (Or, the IQR = 75.5 − 37 = 38.5 deaths, if the median is excluded from both halves of the ordered list.)

## 57. Floods 2105 II.

The distribution of deaths from floods is slightly skewed to the right and bimodal. There is one mode at about 40 deaths and one at about 80 deaths. There is one extreme value at 180 deaths.

**58. Tornadoes 2015 II.**

The distribution of deaths from tornadoes is slightly skewed to the right, with one extreme outlier at 553. The median is 54 deaths, and the IQR is 32 deaths.

**59. Pizza prices.**

The mean and standard deviation would be used to summarize the distribution of pizza prices, since the distribution is unimodal and symmetric.

**60. Neck size.**

The mean and standard deviation would be used to summarize the distribution of neck sizes, since the distribution is unimodal and symmetric.

**61. Pizza prices again.**

a) The mean pizza price is closest to $2.60. That's the balancing point of the histogram.

b) The standard deviation in pizza prices is closest to $0.15, since that is the typical distance to the mean. There are no pizza prices as far as $0.50 or $1.00.

**62. Neck sizes again.**

a) The mean neck size is closest to 15 inches. That's the balancing point of the histogram.

b) The standard deviation in neck sizes is closest to 1 inch, because a typical value lies about 1 inch from the mean. There are a few points as far away as 3 inches from the mean, and none as far away as 5 inches. Those are too large to be the standard deviation.

**63. Movie lengths 2010.**

a) A typical movie would be around 105 minutes long. This is near the center of the unimodal and slightly skewed histogram, with the outlier set aside.

b) You would be surprised to find that your movie ran for 150 minutes. Only 3 movies ran that long.

c) The mean run time would probably be higher, since the distribution of run times is skewed to the right, and also has a high outlier. The mean is pulled towards this tail, while the median is more resistant. However, it is difficult to predict what the effect of the low outlier might be from just looking at the histogram.

**64. Golf drives 2015.**

a) The distribution of golf drives is roughly unimodal and symmetric, with a typical drive of around 290 yards. Professional golfers on the men's PGA tour had drives that were as short as about 260 yards, and as long as about 320 yards.

b) Approximately 25% of professional male golfers drive less than 280 yards.

c) The actual mean drive is about 288.69 yards, so any estimate between 285 and 290 yards is reasonable.

d) The distribution of golf drives is approximately symmetric, so the mean and the median should be relatively close. The actual median is 288.7.

**65. Movie lengths 2010 II.**

a) i)  The distribution of movie running times is fairly consistent, with the middle 50% of running times between 98 and 116 minutes.  The interquartile range is 18 minutes.

ii)  The standard deviation of the distribution of movie running times is 16.6 minutes, which indicates that movies typically varied from the mean running time by 16.6 minutes.

b)  Since the distribution of movie running times is skewed to the right and contains an outlier, the standard deviation is a poor choice of numerical summary for the spread.  The interquartile range is better, since it is resistant to outliers.

**66. Golf drives II.**

a) i)  The distribution of PGA golf drives is fairly consistent, with the middle 50% of the drives having distances between 282 and 294.5 yards.  The interquartile range is 12.5 yards.

ii)  The standard deviation of the distribution of PGA golf drives is 9.8 yards, which indicates that golf drives are typically within 9.8 yards of the mean gold drive.

b)  Since the distribution of golf drives is reasonably symmetric, both the standard deviation and the interquartile range are reasonable measures of spread.

**67. Movie budgets.**

The industry publication is using the median, while the watchdog group is using the mean.  It is likely that the mean is pulled higher by a few very expensive movies.

**68. Cold weather.**

a)  The mean temperature will be lower.  The median temperature will not change, since the incorrect temperature is still the lowest temperature, and the median is based only on position.

b)  The range and standard deviation in temperature will both increase, since the incorrect temperature is more extreme than the correct temperature.  The IQR will not change, since the both the correct and incorrect scores are below the first quartile, and the IQR measures the distance between the first and third quartiles.

**69. Gasoline 2014.**

a)
```
            Gasoline Prices

    31 |1
    31 |5
    32 |1233
    32 |6678
    33 |
    33 |9
    34 |23
    34 |556
    Key : 32 | 1 = $3.21/gal
```

**b)** The distribution of gas prices is bimodal, with two clusters, one centered around $3.45 per gallon, and another centered around $3.25 per gallon.  The lowest and highest prices were $3.11 and $3.46 per gallon.

**c)** There is a gap in the distribution of gasoline prices.  There were no stations that charged between $3.28 and $3.39.

**70. The great one.**

**a)**

```
Wayne Gretzsky –
Games played per season
  8 | 000000122
  7 | 8899
  7 | 0344
  6 |
  6 | 4          Key:
  5 |              7 | 8  =  78
  5 |                      games
  4 | 58
  4 |
```

**b)** The distribution of the number of games played by Wayne Gretzky is unimodal and skewed to the left.

**c)** Typically, Wayne Gretzky played about 80 games per season.  The number of games played is tightly clustered in the upper 70s and low 80s.

**d)** Two seasons are low outliers, when Gretzky played fewer than 50 games.  He may have been injured during those seasons.  Regardless of any possible reasons, these seasons were unusual compared to Gretzky's other seasons.

**71. States.**

**a)** There are 50 entries in the stemplot, so the median must be between the 25th and 26th population values.  Counting in the ordered stemplot gives median = 4.5 million people.  The middle of the lower 50% of the list  (25 state populations) is the 13th population, or 2 million people.  The middle of the upper half of the list (25 state populations) is the 13th population from the top, or 7 million people.  The IQR = Q3 – Q1 = 7 – 2 = 5 million people.

**b)** The distribution of population for the 50 U.S. States is unimodal and skewed heavily to the right.  The median population is 4.5 million people, with 50% of states having populations between 2 and 7 million people.  There are two outliers, a state with 37 million people, and a state with 25 million people.  The next highest population is only 19 million.

**72. Wayne Gretzky.**

**a)** The distribution of the number of games played per season by Wayne Gretzky is skewed to the left, and has low outliers.  The median is more resistant to the skewness and outliers than the mean.

**b)** The median, or middle of the ordered list, is 79 games. Both the 10th and 11th values are 79, so the median is the average of these two, also 79.

**c)** The mean should be lower. There are two seasons when Gretzky played an unusually low number of games. Those seasons will pull the mean down.

## 73. A-Rod 2016.

The distribution of the number of homeruns hit by Alex Rodriguez during the 1994 – 2016 seasons is reasonably symmetric, with the exception of a second mode around 10 homeruns. A typical number of homeruns per season was in the high 30s to low 40s. With the exception of 5 seasons in which A-Rod hit 0, 0 , 5, 7, and 9 homeruns, his total number of homeruns per season was between 16 and the maximum of 57.

## 74. Major hurricanes 2013.

**a)** A dotplot of the number of hurricanes each year from 1944 through 2013 is displayed. Each dot represents a year in which there were that many hurricanes.

**b)** The distribution of the number of hurricanes per year is unimodal and skewed to the right, with center around 2 hurricanes per year. The number of hurricanes per year ranges from 0 to 8. There are no outliers. There may be a second mode at 5 hurricanes per year, but since there were only 6 years in which 5 hurricanes occurred, this may simply be natural variability.

## 75. A-Rod again 2016.

**a)** This is not a histogram. The horizontal axis should contain the number of home runs per year, split into bins of a convenient width. The vertical axis should show the frequency; that is, the number of years in which A-Rod hit a number of home runs within the interval of each bin. The display shown is a bar chart/time plot hybrid that simply displays the data table visually. It is of no use in describing the shape, center, spread, or unusual features of the distribution of home runs hit per year by A-Rod.

**b)** The histogram is at the right.

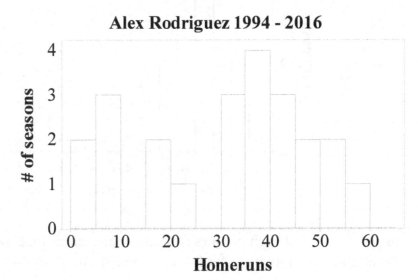

**76. Return of the birds 2013.**

**a)** This is not a histogram.  The horizontal axis should split the number of counts from each site into bins.  The vertical axis should show the number of sites in each bin.  The given graph is nothing more than a bar chart, showing the bird count from each site as its own bar.  It is of absolutely no use for describing the shape, center, spread, or unusual features of the distribution of bird counts.

**b)** The histogram is at the right.

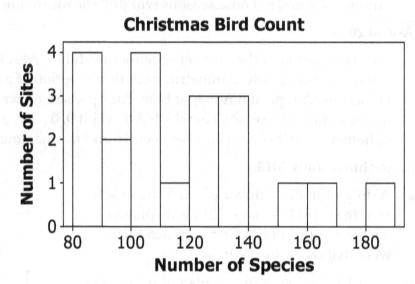

**77. Acid rain.**

**a)** The distribution of the pH readings of water samples in Allegheny County, Penn. is bimodal.  A roughly uniform cluster is centered around a pH of 4.4.  This cluster ranges from pH of 4.1 to 4.9.  Another smaller, tightly packed cluster is centered around a pH of 5.6.  Two readings in the middle seem to belong to neither cluster.

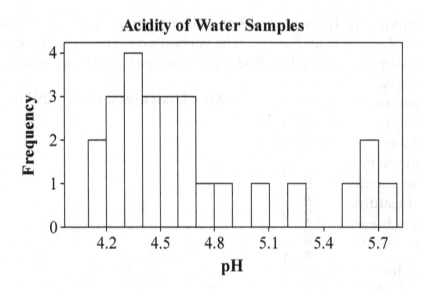

**b)** The cluster of high outliers contains many dates that were holidays in 1973. Traffic patterns would probably be different then, which might account for the difference.

## 78. Marijuana 2015.

The distribution of the percentage of 16-year-olds in 38 countries who have used marijuana in 2015 is somewhat bimodal, with a group of countries having approximately 5 to 10% of 16-year-olds having used marijuana. Another group of countries has between 15% and 25% of teens who have used marijuana. Kosovo, at 2%, had the lowest percentage of 16-year-olds who have tried marijuana. Czech Republic had the highest percentage, at 42%. A typical country might have a percentage of approximately 18%, the median percentage of marijuana use among 16-year-olds.

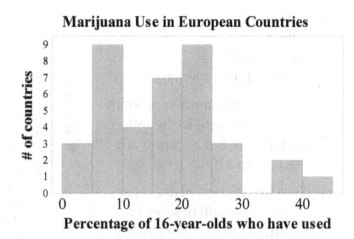

**Marijuana Use in European Countries**

Percentage of 16-year-olds who have used

## 79. Final grades.

The width of the bars is much too wide to be of much use. The distribution of grades is skewed to the left, but not much more information can be gathered.

## 80. Final grades revisited.

a) This display has a bar width that is much too narrow. As it is, the histogram is only slightly more useful than a list of scores. It does little to summarize the distribution of final exam scores.

b) The distribution of test scores is skewed to the left, with center at approximately 170 points. There are several low outliers below 100 points, but other than that, the distribution of scores is fairly tightly clustered.

## 81. Zip codes.

Even though zip codes are numbers, they are not quantitative in nature. Zip codes are categories. A histogram is not an appropriate display for categorical data. The histogram the Holes R Us staff member displayed doesn't take into account that some 5-digit numbers do not correspond to zip codes or that zip codes falling into the same classes may not even represent similar cities or towns. The employee could design a better display by constructing a bar chart that groups together zip codes representing areas with similar demographics and geographic locations.

## 82. Zip codes revisited

The statistics cannot tell us very much since zip codes are categorical. However, there is *some* information in the first digit of zip codes. They indicate a general East (0-1) to West (8-9) direction. So, the distribution shows that a large portion of their sales occurs in the West and another in the 32000 area. But a bar chart of the first digits would be the appropriate display to show this information.

### 83. Math scores 2013.

a) Median: 285
   IQR: 9
   Mean: 284.36
   Standard deviation: 6.84

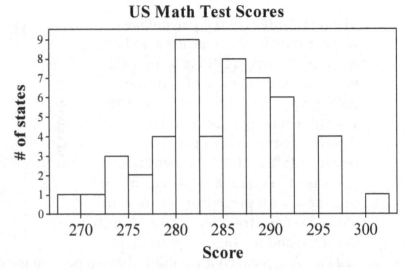

b) Since the distribution of Math scores is skewed to the left, it is probably better to report the median and IQR.

c) The distribution of average math achievement scores for eighth graders in the United States is skewed slightly to the left, and roughly unimodal. The distribution is centered at 285. Scores range from 269 to 301, with the middle 50% of the scores falling between 280 and 289.

### 84. Boomtowns 2015.

a) A histogram of the job growth rates of NewGeography.com's best cities for job growth is at the right. A boxplot, stemplot, or dotplot would also have been an acceptable display.

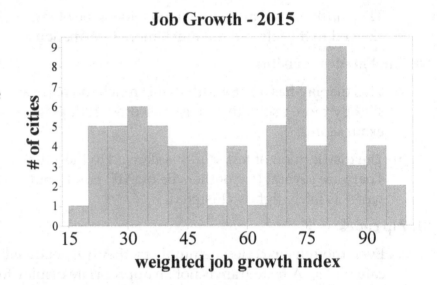

b) The mean weighted job rating index is 58.26% and the median weighted job rating index is 61.70%. The mean is lower than the median, indicating that the distribution is skewed to the left. At the very least, the distribution is bimodal, which makes the mean and median difficult to predict without calculations.

c) It might be more informative to report one mode at about 33% and the other mode at about 80%.

d) The standard deviation of the distribution of weighted job rating indices is 23.85% and the IQR is 45.1%.

e) Neither is ideal. It might be more informative to discuss a measure of spread for each mode.

f) If 49.23% were subtracted from each of the weighted job rating indices, the mean and median would each decrease by 49.23%. The standard deviation and the IQR would not change.

g) If we were to set aside Austin-Round Rock, the 4th-highest weighted job rating index, the mean and standard deviation would decrease. The median and IQR would be relatively unaffected, although they would change slightly, since they are each based upon relative position. With the 4th-highest rating removed, there would only be 69 rating indices, instead of 70. This would cause the median and the quartiles to shift down slightly.

h) The distribution of weighted job rating indices is bimodal and slightly skewed to the left. The lower mode is centered at a weighted job rating index of approximately 33%, and the upper mode is centered at approximately 80%.

**85. Population growth 2010.**

The distribution of population growth among the 50 United States and the District of Columbia is unimodal and skewed to the right. Most states experienced modest growth, as measured by percent change in population between 2000 and 2010. Nearly every state experienced positive growth, with the exception of Michigan. The median population growth was 7.8%, with the middle 50% of states experiencing between 4.30% and 14.10% growth, for an IQR of 9.80. The distribution contains one high outlier. Nevada experienced population growth of 35.1%.

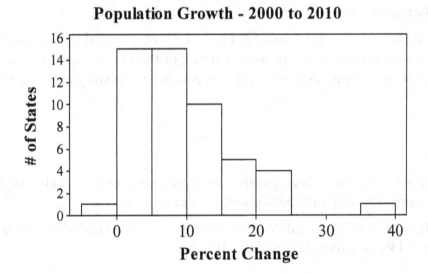

**86. Student survey.**

Answers will vary. The best answers are about students and their responses, not about numbers and displays. Write what you learn about students, not about what your display looks like. Don't give the value of a summary statistic without a discussion of what it might mean.

## Chapter 3 – Relationships Between Categorical Variables – Contingency Tables

### Section 3.1

**1. College value?**

|  | Poor | Only Fair | Good | Excellent | DK/NA | Total |
|---|---|---|---|---|---|---|
| US Adults | 321 | 900 | 750 | 107 | 64 | 2142 |
| Presidents | 32 | 222 | 622 | 179 | 0 | 1055 |
| Total | 353 | 1122 | 1372 | 286 | 64 | 3197 |

a) The percent of college presidents who think that higher education provides a poor value is $32/1055 \approx 3\%$.

b) $(750+107)/2142 \approx 40\%$ of U.S. adults think that the value provided by the higher education system is either good or excellent.

c) 15% of U.S. adults view college as a poor value, but only 3% of college presidents do. Similarly, U.S. adults are twice as likely to view college as an only fair value compared to the presidents (42% to 21%). Presidents are much more likely (76%) to rate colleges as a good or excellent value compared to U.S. adults (only 40%). So in short, college presidents have a much higher opinion of the value of college than U.S. adults do.

d) In this random sample of U.S. adults, $107/2142 \approx 5.00\%$ responded that college provides an excellent value. This is only an estimate, however, of the percentage of all U.S. adults who feel the same way. The percentage of all U.S. adults is probably close to, but not exactly 5%.

### Section 3.2

**3. College value again.**

a) The conditional distribution of college presidents' opinions about the value of a college education is 3% poor; 21% only fair; 59% good; 17% excellent.

b) Omitting the 64 DK/NAs, the conditional distribution of the opinions U.S. adults about the value of a college education is 15% negative; 43% middle; 41% positive.

### Section 3.3

**5. Diet and politics.**

a) The distribution of political alignment among carnivores is about 25% conservative, 40% moderate, and 35% liberal. Omnivores were generally less conservative and more liberal, with about 10% conservative, 35% moderate, and 55% liberal. Vegetarians were even more liberal than the other groups, with 0% conservative, 30% moderate, and 70% liberal.

b) The differences are quite large. There appears to be a strong association between diet preference and political alignment.

7. **Fish and prostate cancer revisited**

   a) Looking at the horizontal axis only, approximately 7% of the men had prostate cancer.

   b) There are more men who didn't have cancer and never or seldom ate fish. The rectangles are approximately the same height, but the bar for "no cancer" is much wider.

   c) The percentage of men who never/seldom ate fish is lower in the group with no cancer than in the group with cancer. Disregard the width, and look only at the height to compare the conditional distribution of fish consumption within each cancer group. The bar for "never/seldom" is slightly shorter within the "no cancer" group.

**Section 3.4**

9. **Diet and politics III.**

   **Men**

   |  | Carnivore | Omnivore | Vegetarian | Total |
   |---|---|---|---|---|
   | Liberal | 9 | 74 | 5 | 88 |
   | Moderate | 12 | 54 | 1 | 67 |
   | Conservative | 9 | 14 | 0 | 23 |
   | Total | 30 | 142 | 6 | 178 |

   **Women**

   |  | Carnivore | Omnivore | Vegetarian | Total |
   |---|---|---|---|---|
   | Liberal | 4 | 53 | 12 | 69 |
   | Moderate | 4 | 27 | 6 | 37 |
   | Conservative | 1 | 4 | 0 | 5 |
   | Total | 9 | 84 | 18 | 111 |

   a) Men are more likely to be conservative carnivores. $9/178 \approx 5.1\%$ of the men are conservative carnivores, while only $1/111 \approx 0.9\%$ of the women are conservative carnivores.

   b) Liberal vegetarians are more likely to be women. Of the 17 liberal vegetarians, 12 of them are women. $12/17 \approx 70.6\%$ of liberal vegetarians are women.

**Chapter Exercises**

11. **Movie genres and ratings.**

   a) $452/1529 \approx 29.6\%$ of the films were rated R.

   b) $124/1529 \approx 8.1\%$ of the films were R-rated comedies.

   c) $124/452 \approx 27.4\%$ of the R-rated films were comedies.

   d) $124/312 \approx 39.7\%$ of the comedies were rated R.

**13. Tables in the news.** Answers will vary.

**15. Poverty and region 2012.**

The differences in poverty are not huge, but they may be real. The Northeast and Midwest have the lowest percentages of people living below the poverty level: 12.7% and 13.7%, respectively. In the West, 15.4% live below the poverty level, and the South has the highest rate at 16.8%.

**17. Death from the sky.**

a) $100 - 60 - 30 - 1 - 0.2 - 0.17 = 8.63\%$ of estimated deaths are attributed to causes not listed here.

b) Regardless of the type of display chosen, it is difficult to display causes of death with percentages as low as 0.2% and 0.17%.

**19. Smoking.**

a) The smoking rate for 18-24-year-old men was 42.1% in 1974.

b) From 1974 to 2014, the smoking rate for 18-24-year-old men dropped from 42.1% to 18.5%

c) Men who were 18-24 years old in 1974 are in the 35-44 age group in 1994, the 45-54 age group in 2004, and the 2014. The smoking rate for this cohort has been decreasing through the years, from 42.1% to 33.2% to 26.7% to 18.8%. Although we don't have data on deaths in this table, it may very well be that the smokers have a higher death rate than the non-smokers, so this decrease doesn't necessarily mean that men in this cohort are quitting smoking.

**21. Mothers and fathers 1965-2011.**

a) Fathers spend the vast majority of their time on paid work, while mothers spend more time on child care and house work.

b) The time fathers spend on paid work has decreased, and the time they spend on child care and housework has increased. For mothers, the number of hours spent on paid work has significantly increased, and they have also increased their time spent on child care while reducing housework time.

c) Parents are spending more time on child care and paid work (13 hours to 21 hours and 50 hours to 58 hours). The time spent on housework has decreased from 36 hours to 28 hours.

d) Overall, parents in 2011 reported spending more time total on these tasks, a total of 107 hours in 2011 compared to 99 hours in 1965. Mothers increased their total working time by 3 hours, from 50 hours to 53 hours, while fathers increased their total working time by 5 hours, from 49 hours to 54 hours.

## 23. Teen smokers.

According to the Monitoring the Future study, teen smoking brand preferences differ somewhat by region. Although Marlboro is the most popular brand in each region, with about 58% of teen smokers preferring this brand in each region, teen smokers from the South prefer Newports at a higher percentage than teen smokers from the West, 22.5% to approximately 10%, respectively. Camels are more popular in the West, with 9.5% of teen smokers preferring this brand, compared to only 3.3% in the South. Teen smokers in the West are also more likely to have no particular brand than teen smokers in the South. 12.9% of teen smokers in the West have no particular brand, compared to only 6.7% in the South. Both regions have about 9% of teen smokers that prefer one of over 20 other brands.

## 25. Diet and politics IV.

a) There are more men in the survey. The male columns are generally wider than the female columns.

b) We can't compare the genders within each category of political ideology since the sample sizes differ. We can, however, note that the male bars are narrowest in the Liberal category, and widest in the Conservative category, indicating that a small share of liberals and a large share of conservatives among the men. The women show the opposite association. The widest female bar is Liberal, while the narrowest is Conservative. In other words, there is an association between gender and political ideology. Males tend to be more conservative and females tend to be more liberal.

c) There is an association between politics and diet. Conservatives are more likely to be carnivores, while liberals are more likely to be vegetarians.

d) The association between politics and diet seems to differ between men and women. Differences in vegetarianism across political ideology is more pronounces in females than in males. Differences in carnivorous eating habits across political ideology is more pronounced in males than females.

## 27. Job satisfaction.

a) This is a table of column percents. The columns add up to 100%, while the rows do not.

b) i) This can't be found from the table. We don't know what the percent of respondents who are very satisfied.

ii) This can't be found from the table. We don't know what the percent of respondents who are dissatisfied.

iii) 39% of respondents who are dissatisfied with their current job are actually better off than their parents were at the same age.

iv) This can't be found from the table. We don't know what the percent of respondents who are very satisfied.

**29. Seniors.**

**a)** A table with marginal totals is to the right. There are 268 White graduates and 325 total graduates. $268/325 \approx 82.5\%$ of the graduates are white.

| Plans | White | Minority | TOTAL |
|---|---|---|---|
| 4-year college | 198 | 44 | 242 |
| 2-year college | 36 | 6 | 42 |
| Military | 4 | 1 | 5 |
| Employment | 14 | 3 | 17 |
| Other | 16 | 3 | 19 |
| **TOTAL** | 268 | 57 | 325 |

**b)** There are 42 graduates planning to attend 2-year colleges. $42/325 \approx 12.9\%$

**c)** 36 white graduates are planning to attend 2-year colleges. $36/325 \approx 11.1\%$

**d)** 36 white graduates are planning to attend 2-year colleges and there are 268 whites graduates. $36/268 \approx 13.4\%$

**e)** There are 42 graduates planning to attend 2-year colleges, and 36 of them are white. $36/42 \approx 85.7\%$

**31. Movies 06-15.**

**a)** This is a table of column percents. The columns add up to 100%, while the rows do not.

**b)** Movies rated G and PG have become slightly less common, while movies rated PG-13 and R have become slightly more common.

**c)** For Dramas, the percentages of PG and R have decreased while the percentage of PG-13 dramas has increased significantly. For Comedies, there has been a large increase in the percentage of R-rated films.

**33. More about seniors.**

**a)** For white students, 73.9% plan to attend a 4-year college, 13.4% plan to attend a 2-year college, 1.5% plan on the military, 5.2% plan to be employed, and 6.0% have other plans.

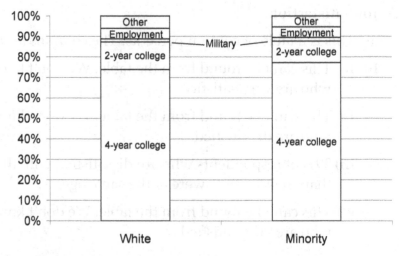

**b)** For minority students, 77.2% plan to attend a 4-year college, 10.5% plan to attend a 2-year college, 1.8% plan on the military, 5.3% plan to be employed, and 5.3% have other plans.

**c)** A segmented bar chart is a good display of these data.

**d)** The conditional distributions of plans for Whites and Minorities are similar: White – 74% 4-year college, 13% 2-year college, 2% military, 5% employment, 6% other. Minority – 77% 4-year college, 11% 2-year college, 2% military, 5% employment, 5% other. Caution should be used with the percentages for Minority graduates, because the total is so small. Each graduate is almost 2%. Still, the conditional distributions of plans are essentially the same for the two groups. There is little evidence of an association between race and plans for after graduation.

**35. Magnet schools revisited.**

**a)** There were 1755 qualified applicants to the Houston Independent School District's magnet schools program. Of those, 292, or about 16.6% were Asian.

**b)** There were 931 students accepted to the magnet schools program. Of those, 110, or about 11.8% were Asian.

**c)** There were 292 Asian applicants. Of those, 110, or about 37.7%, were accepted.

**d)** There were 1755 total applicants. Of those, 931, or about 53%, were accepted.

**37. Back to school.**

There were 1,755 qualified applicants for admission to the magnet schools program. 53% were accepted, 17% were wait-listed, and the other 30% were turned away. While the overall acceptance rate was 53%, 93.8% of Blacks and Hispanics were accepted, compared to only 37.7% of Asians, and 35.5% of whites. Overall, 29.5% of applicants were Black or Hispanics, but only 6% of those turned away were Black or Hispanic. Asians accounted for 16.6% of applicants, but 25.3% of those turned away. It appears that the admissions decisions were not independent of the applicant's ethnicity.

**39. Weather forecasts.**

**a)** The table shows the marginal totals. It rained on 34 of 365 days, or 9.3% of the days.

**b)** Rain was predicted on 90 of 365 days. $90/365 \approx 24.7\%$ of the days.

|  |  | Actual Weather | | |
|---|---|---|---|---|
|  |  | Rain | No Rain | **Total** |
| **Forecast** | Rain | 27 | 63 | 90 |
|  | No Rain | 7 | 268 | 275 |
|  | **Total** | 34 | 331 | 365 |

**c)** The forecast of Rain was correct on 27 of the days it actually rained and the forecast of No Rain was correct on 268 of the days it didn't rain. So, the forecast was correct a total of 295 times. $295/365 \approx 80.8\%$ of the days.

**d)** On rainy days, rain had been predicted 27 out of 34 times (79.4%). On days when it did not rain, forecasters were correct in their predictions 268 out of 331 times (81.0%). These two percentages are very close. There is no evidence of an association between the type of weather and the ability of the forecasters to make an accurate prediction.

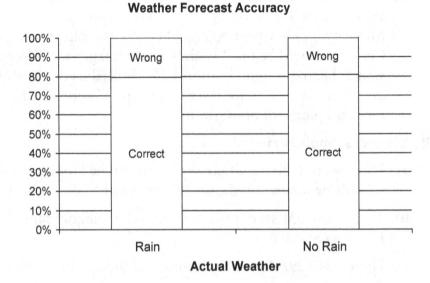

**41. Blood pressure.**

**a)** The marginal distribution of blood pressure for the employees of the company is the total column of the table, converted to percentages. 20% low, 49% normal and 31% high blood pressure.

| Blood pressure | under 30 | 30 - 49 | over 50 | Total |
|---|---|---|---|---|
| low | 27 | 37 | 31 | 95 |
| normal | 48 | 91 | 93 | 232 |
| high | 23 | 51 | 73 | 147 |
| Total | 98 | 179 | 197 | 474 |

**b)** The conditional distribution of blood pressure within each age category is:
Under 30 : 28% low, 49% normal, 23% high
30 – 49 : 21% low, 51% normal, 28% high
Over 50 : 16% low, 47% normal, 37% high

**c)** A segmented bar chart of the conditional distributions of blood pressure by age category is at the right.

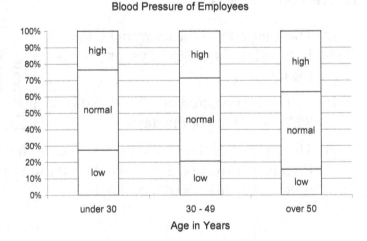

**d)** In this company, as age increases, the percentage of employees with low blood pressure decreases, and the percentage of employees with high blood pressure increases.

**e)** No, this does not prove that people's blood pressure increases as they age. Generally, an association between two variables does not imply a cause-and-effect relationship. Specifically, these data come from only one company and cannot be applied to all people. Furthermore, there may be some other variable that is linked to both age and blood pressure. Only a controlled experiment can isolate the relationship between age and blood pressure.

**43. Anorexia.**

These data provide no evidence that Prozac might be helpful in treating anorexia. About 71% of the patients who took Prozac were diagnosed as "Healthy", while about 73% of the patients who took a placebo were diagnosed as "Healthy". Even though the percentage was higher for the placebo patients, this does not mean that Prozac is hurting patients. The difference between 71% and 73% is not likely to be statistically significant.

**45. Driver's licenses 2014.**

a) There are 8.5 million drivers under 20 and a total of 214.1 million drivers in the U.S. That's about 4.0% of U.S. drivers under 20.

b) There are 105.9 million males out of 214.1 million total U.S. drivers, or about 49.5%.

c) Each age category appears to have about 50% male and 50% female drivers. At younger ages, males form the slight majority of drivers. The percentage of male drivers shrinks until female drivers hold a slight majority, among older drivers.

d) There appears to be a slight association between age and gender of U.S. drivers. Younger drivers are slightly more likely to be male, and older drivers are slightly more likely to be female.

**47. Diet and politics shuffled.**

a) The upper right bar chart shows the original data.

b) It's clear that shuffled data did not produce associations as obvious as the original data. The randomly scrambled data look different from the original data, and this supports the belief that there is an association between the variables.

**49. Hospitals.**

a) The marginal totals have been added to the table:

| | | \multicolumn{3}{c}{Discharge delayed} | | |
|---|---|---|---|---|
| | | Large Hospital | Small Hospital | Total |
| Procedure | Major surgery | 120 of 800 | 10 of 50 | 130 of 850 |
| | Minor surgery | 10 of 200 | 20 of 250 | 30 of 450 |
| | Total | 130 of 1000 | 30 of 300 | 160 of 1300 |

160 of 1300, or about 12.3% of the patients had a delayed discharge.

b) Yes. Major surgery patients were delayed 130 of 850 times, or about 15.3% of the time. Minor Surgery patients were delayed 30 of 450 times, or about 6.7% of the time.

c) Large Hospital had a delay rate of 130 of 1000, or 13%.
Small Hospital had a delay rate of 30 of 300, or 10%.
The small hospital has the lower overall rate of delayed discharge.

**d)** Large Hospital: Major Surgery 15% delayed and Minor Surgery 5% delayed. Small Hospital: Major Surgery 20% delayed and Minor Surgery 8% delayed. Even though small hospital had the lower overall rate of delayed discharge, the large hospital had a lower rate of delayed discharge for each type of surgery.

**e)** No. While the overall rate of delayed discharge is lower for the small hospital, the large hospital did better with *both* major surgery and minor surgery.

**f)** The small hospital performs a higher percentage of minor surgeries than major surgeries. 250 of 300 surgeries at the small hospital were minor (83%). Only 200 of the large hospital's 1000 surgeries were minor (20%). Minor surgery had a lower delay rate than major surgery (6.7% to 15.3%), so the small hospital's overall rate was artificially inflated. Simply put, it is a mistake to look at the overall percentages. The real truth is found by looking at the rates after the information is broken down by type of surgery, since the delay rates for each type of surgery are so different. The larger hospital is the better hospital when comparing discharge delay rates.

## 51. Graduate admissions.

**a)** 1284 applicants were admitted out of a total of 3014 applicants. 1284/3014 = 42.6%

| Program | Males Accepted (of applicants) | Females Accepted (of applicants) | Total |
|---------|-------------------------------|----------------------------------|-------|
| 1 | 511 of 825 | 89 of 108 | 600 of 933 |
| 2 | 352 of 560 | 17 of 25 | 369 of 585 |
| 3 | 137 of 407 | 132 of 375 | 269 of 782 |
| 4 | 22 of 373 | 24 of 341 | 46 of 714 |
| **Total** | **1022 of 2165** | **262 of 849** | **1284 of 3014** |

**b)** 1022 of 2165 (47.2%) of males were admitted. 262 of 849 (30.9%) of females were admitted.

**c)** Since there are four comparisons to make, the table at the right organizes the percentages of males and females accepted in each program. Females are accepted at a higher rate in every program.

| Program | Males | Females |
|---------|-------|---------|
| 1 | 61.9% | 82.4% |
| 2 | 62.9% | 68.0% |
| 3 | 33.7% | 35.2% |
| 4 | 5.9% | 7% |

**d)** The comparison of acceptance rate within each program is most valid. The overall percentage is an unfair average. It fails to take the different numbers of applicants and different acceptance rates of each program. Women tended to apply to the programs in which gaining acceptance was difficult for everyone. This is an example of Simpson's Paradox.

# Chapter 4 – Understanding and Comparing Distributions

**Section 4.1**

**1. Load factors, 2016.**

The distribution of domestic load factors and the distribution of international load factors are both unimodal and skewed to the left. The distribution of international load factors may contain a low outlier. Because the distributions are skewed, the median and IQR are the appropriate measures of center and spread. The medians are very close, which tell us that typical international and domestic load factors are about the same. The IQRs show a bit more variability in the domestic load factors.

**3. Fuel economy.**

If the Mitsubishi i-MiEV is removed, the standard deviation of fuel economy ratings would be much lower, dropping from 11.87 miles per gallon to 4.6 miles per gallon. The IQR would be affected very little, if at all.

**5. Load factors 2016 by month.**

Load factors are generally higher and less variable in the summer months (June – August). They are lower and more variable in the winter and spring. There are several months with low outliers and October has a far low outlier that is the minimum of all the data.

**Section 4.2**

**7. Extraordinary months.**

Air travel immediately after the events of 9/11 was not typical of air travel in general. If we want to analyze monthly patterns, it might be best to set these months aside.

**Section 4.3**

**9. Exoplanets.**

It is difficult to summarize data with a distribution this skewed. The extremely large values will dominate any summary or description.

**Chapter Exercises**

**11. In the news.** Answers will vary.

**13. Pizza prices by city.**

a) Pizza prices appear to be both higher on average, and more variable, in Baltimore than in the other three cities. Prices in Chicago may be slightly higher on average than in Dallas and Denver, but the difference is small.

**b)** There are low outliers in the distribution of pizza prices in Baltimore and Chicago.  There is one high outlier in the distribution of pizza prices in Dallas. These outliers do not affect the overall conclusions reached in the previous part.

**15. Cost of living 2016, selected cities.**

**a)** Cappuccino is the most expensive commodity, in general. The first quartile of cappuccino prices is higher than all the water prices. The middle 50%of cappuccino prices is less variable than the middle 50% of egg prices. The third quartile and maximum value of cappuccino prices are both higher than those for egg prices.

**b)** We can't say anything about price comparisons in individual cities from this display. The minimum price of a cappuccino is lower than the minimum price of a carton of eggs. It could be the case that those two prices are in the same city, and cappuccino costs less than eggs.

**17. Rock concert deaths.**

**a)** The histogram and boxplot of the distribution of "crowd crush" victims' ages both show that a typical crowd crush victim was approximately 18 - 20 years of age, that the range of ages is 36 years, that there are two outliers, one victim at age 36 - 38 and another victim at age 46 – 48.

**b)** This histogram shows that there may have been two modes in the distribution of ages of "crowd crush" victims, one at 18 - 20 years of age and another at 22 – 24 years of age.  Boxplots, in general, can show symmetry and skewness, but not features of shape like bimodality or uniformity. Most victims were between 16 and 24 years of age.

**c)** The median is the better measure of center, since the distribution of ages has outliers.  Median is more resistant to outliers than the mean.

**d)** The IQR is a better measure of spread, since the distribution of ages has outliers. IQR is more resistant to outliers than the standard deviation.

**19. Sugar in cereals.**

**a)** The maximum sugar content is approximately 60% and the minimum sugar content is approximately 1%, so the range of sugar contents is about 60 – 1 = 59%.

**b)** The distribution of sugar content of cereals is bimodal, with modes centered around 5% and 45% sugar by weight.

**c)** Some cereals are healthy, low-sugar brands, and others are very sugary.

**d)** Yes.  The minimum sugar content in the children's cereals is about 35% and the maximum sugar content of adult cereals is only 34%.

e) The range of sugar contents is about the same for the two types of cereals, approximately 28%, but the IQR is larger for the adult cereals. This is an indication of more variability in the sugar content of the middle 50% of adult cereals.

**21. Population growth 2010 by region.**

a) Comparative boxplots are at the right.

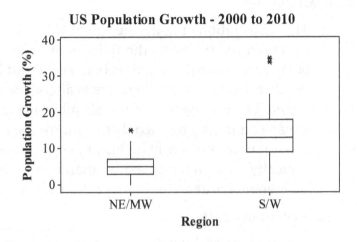

b) The distribution of population growth in NE/MW states is unimodal, symmetric and tightly clustered around 5% growth. The distribution of population growth in S/W states is much more spread out, with most states having population growth between 5% and 25%. A typical state had about 15% growth. There were two outliers, with 34% and 35% growth, respectively. Generally, the growth rates in the S/W states were higher and more variable than the rates in the NE/MW states.

**23. Hospital stays.**

a) The histograms of male and female hospital stay durations would be easier to compare if the were constructed with the same scale, perhaps from 0 to 20 days.

b) The distribution of hospital stays for men is skewed to the right, with many men having very short stays of about 1 or 2 days. The distribution tapers off to a maximum stay of approximately 25 days. The distribution of hospital stays for women is skewed to the right, with a mode at approximately 5 days, and tapering off to a maximum stay of approximately 22 days. Typically, hospital stays for women are longer than those for men.

c) The peak in the distribution of women's hospital stays can be explained by childbirth. This time in the hospital increases the length of a typical stay for women, and not for men.

**25. Women's basketball.**

a) Both girls have a median score of about 17 points per game, but Scyrine is much more consistent. Her IQR is about 2 points, while Alexandra's is over 10.

**b)** If the coach wants a consistent performer, she should take Scyrine. She'll almost certainly deliver somewhere between 15 and 20 points. But, if she wants to take a chance and needs a "big game", she should take Alexandra. Alex scores over 24 points about a quarter of the time. On the other hand, she scores under 11 points about as often.

**27. Marriage age.**

The distribution of marriage age of U.S. men is skewed right, with a typical man (as measured by the median) first marrying at around 24 years old. The middle 50% of male marriage ages is between about 23 and 26 years. For U.S. women, the distribution of marriage age is also skewed right, with median of around 21 years. The middle 50% of female marriage age is between about 20 and 23 years. When comparing the two distributions, the most striking feature is that the distributions are nearly identical in spread, but have different centers. Females typically seem to marry earlier than males. In fact, between 50% and 75% of the women marry at a younger age than *any* man.

**29. Fuel economy 2016.**

(Note: numerical details may vary.) In general, fuel economy is highest in mid-size cars, lower in SUVs, and lowest in pickup trucks. There are numerous outliers on the high end for mid-size cars. The top 50% of cars get higher fuel economy than 75% of SUVs and all pickups. The distributions of fuel economy for cars and SUVs have approximately the same IQR, about 8 miles per gallon, but cars generally have higher combined fuel economy. Furthermore, there are several cars with much higher fuel economy. Pickups trucks have consistently low fuel economy.

**31. Test scores.**

Class A is Class 1. The median is 60, but has less spread than Class B, which is Class 2. Class C is Class 3, since it's median is higher, which corresponds to the skew to the left.

**33. Graduation?**

**a)** The distribution of the percent of incoming college freshman who graduate on time is roughly symmetric. The mean and the median are reasonably close to one another and the quartiles are approximately the same distance from the median.

**b)** Upper Fence:  Q3+1.5(IQR)=74.75+1.5(74.75-59.15)

$$=74.75+23.4$$

$$=98.15$$

Lower Fence:  Q1-1.5(IQR)=59.15-1.5(74.75-59.15)

$$=59.15-23.4$$

$$=35.75$$

Since the maximum value of the distribution of the percent of incoming freshmen who graduate on time is 87.4% and the upper fence is 98.15%, there are no high outliers. Likewise, since the minimum is 43.2% and the lower fence is 35.75%, there are no low outliers. Since the minimum and maximum percentages are within the fences, all percentages must be within the fences.

**c)** A boxplot of the distribution of the percent of incoming freshmen who graduate on time is at the right.

**d)** The distribution of the percent of incoming freshmen who graduate on time is roughly symmetric, with mean of approximately 68% of freshmen graduating on time. Universities surveyed had between 43.2% and 87.4% of students graduating on time, with the middle 50% of universities reporting between 59.15% and 74.75% graduating on time.

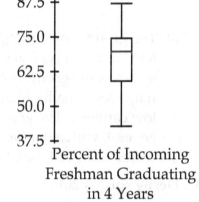

Percent of Incoming Freshman Graduating in 4 Years

## 35. Caffeine.

**a)** *Who* – 45 student volunteeers. *What* – Level of caffeine consumption and memory test score. *When* – Not specified. *Where* – Not specified. *Why* – The student researchers want to see the possible effects of caffeine on memory. *How* – It appears that the researchers imposed the treatment of level of caffeine consumption in an experiment. However, this point is not clear. Perhaps they allowed the subjects to choose their own level of caffeine.

**b)** *Variables* – Caffeine level is a categorical variable with three levels: no caffeine, low caffeine, and high caffeine. Test score is a quantitative variable, measured in number of items recalled correctly.

**c)**

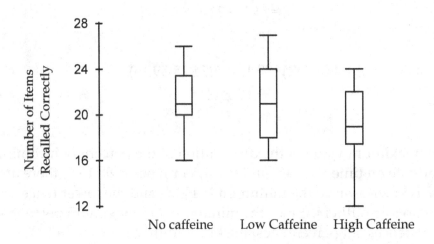

**d)** The groups consuming no caffeine and low caffeine had comparable memory test scores. A typical score from these groups was around 21. However, the scores of the group consuming no caffeine were more consistent, with a smaller range and smaller interquartile range than the scores of the group consuming low caffeine. The group consuming high caffeine had lower memory scores in general, with a median score of about 19. No one in the high caffeine group scored above 24, but 25% of each of the other groups scored above 24.

**37. Derby speeds 2016.**

**a)** The median speed is the speed at which 50% of the winning horses ran slower. Find 50% on the left, move straight over to the graph and down to a speed of about 36 mph.

**Kentucky Derby 2016**

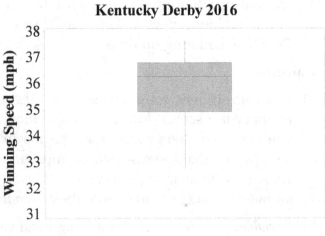

**b)** Quartile 1 is at 25% on the left, and Quartile 3 is at 75% on the left. Matching these to the ogive, Q1 = 35 mph and Q3 = 37 mph, approximately.

**c)** Range = Max – Min = 38 – 31 = 7 mph
IQR = Q3 – Q1 = 36.5 – 34.5 = 2 mph

**d)** A boxplot of winning Kentucky Derby Speeds is at the right.

e) The distribution of winning speeds in the Kentucky Derby is skewed to the left. The lowest winning speed is just under 31 mph and the fastest speed is about 38 mph. The median speed is approximately 36 mph, and 75% of winning speeds are above 35 mph. Only a few percent of winners have had speeds below 33 mph. The middle 50% of winning speeds are between 35 and 37 mph.

## 39. Reading scores.

a) The highest score for boys was 6, which is higher than the highest score for girls, 5.9.

b) The range of scores for boys is greater than the range of scores for girls.
Range = Max – Min       Range(Boys) = 4     Range(Girls) = 3.1

c) The girls had the greater IQR.
IQR = Q3 – Q1       IQR(Boys) = 4.9 – 3.9 = 1          IQR(Girls) = 5.2 – 3.8 = 1.4

d) The distribution of boys' scores is more skewed. The quartiles are not the same distance from the median. In the distribution of girls' scores, Q1 is 0.7 units below the median, while Q3 is 0.7 units above the median.

e) Overall, the girls did better on the reading test. The median, 4.5, was higher than the median for the boys, 4.3. Additionally, the upper quartile score was higher for girls than boys, 5.2 compared to 4.9. The girls' lower quartile score was slightly lower than the boys' lower quartile score, 3.8 compared to 3.9.

f) The overall mean is calculated by weighting each mean by the number of students. $\dfrac{14(4.2)+11(4.6)}{25} = 4.38$

## 41. Industrial experiment.

First of all, there is an extreme outlier in the distribution of distances for the slow speed drilling. One hole was drilled almost an inch away from the center of the target! If that distance is correct, the engineers at the computer production plant should investigate the slow speed drilling process closely. It may be plagued by extreme, intermittent inaccuracy. The outlier in the slow speed drilling process is so extreme that no graphical display can display the distribution in a meaningful way while including that outlier. That distance should be removed before looking at a plot of the drilling distances.

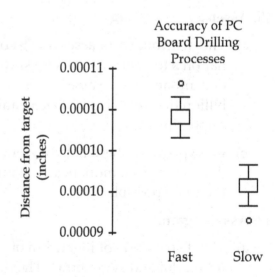

With the outlier set aside, we can determine that the slow drilling process is more accurate. The greatest distance from the target for the slow drilling process, 0.000098 inches, is still more accurate than the smallest distance for the fast drilling process, 0.000100 inches.

**43. MPG.**

a) Comparative boxplots are below. Comparative dotplots, histograms, or stemplots would also be acceptable.

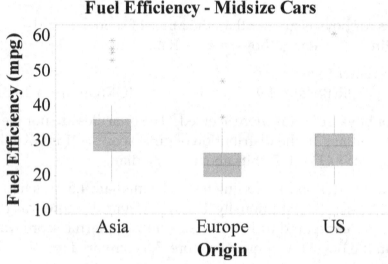

**Fuel Efficiency - Midsize Cars**

b) The distributions of fuel efficiency for all three groups have high outliers, potentially hybrid or electric cars. Asian cars generally have the highest fuel efficiency, while European cars have the lowest. The Asian median is about 32 mpg, European cars around 23 mpg, and U.S. cars 28 mpg.

**45. Assets.**

a) The distribution of assets of 79 companies chosen from the *Forbes* list of the nation's top corporations is skewed so heavily to the right that the vast majority of companies have assets represented in the first bar of the histogram, 0 to 10 billion dollars. This makes meaningful discussion of center and spread impossible.

b) Re-expressing these data by, for example, logs or square roots might help make the distribution more nearly symmetric. Then a meaningful discussion of center might be possible.

**47. Assets again.**

a) The distribution of logarithm of assets is preferable, because it is roughly unimodal and symmetric. The distribution of the square root of assets is still skewed right, with outliers.

**b)** If $\sqrt{Assets} = 50$, then the companies assets are approximately $50^2 = 2500$ million dollars.

**c)** If $\log(Assets) = 3$, then the companies assets are approximately $10^3 = 1000$ million dollars.

**49. Stereograms.**

**a)** The two variables discussed in the description are fusion time and treatment group.

**b)** Fusion time is a quantitative variable, measured in seconds. Treatment group is a categorical variable, with subjects either receiving verbal clues only, or visual and verbal clues.

**c)** Both groups have distributions that are skewed to the right. Generally, the Visual/Verbal group had shorter fusion times than the No/Verbal group. The median for the Visual/Verbal group was approximately the same as the lower quartile for the No/Verbal group. The No/Verbal Group also had an extreme outlier, with at least one subject whose fusion time was approximately 50 seconds. There is evidence that visual information may reduce fusion time.

**51. Cost of living sampled.**

**a)** If there were no mean difference, we would expect the histogram to be centered at 0.

**b)** (Answers may vary.) Because 0 never occurred in 1000 random samples, this provides strong evidence that the difference we saw was not due to chance. We can conclude that there is a real (if small) difference between the mean prices of a cappuccino and a dozen eggs.

## Chapter 5 – The Standard Deviation as a Ruler and the Normal Model

### Section 5.1

**1. Stats test.**

Gregor scored 65 points on the test.

(Or, 75 – 2(5) = 65)

$$z = \frac{y - \mu}{\sigma}$$

$$-2 = \frac{y - 75}{5}$$

$$y = 65$$

**3. Temperatures.**

In January, with mean temperature 36° and standard deviation in temperature 10°, a high temperature of 55° is almost 2 standard deviations above the mean. In July, with mean temperature 74° and standard deviation 8°, a high temperature of 55° is more than two standard deviations below the mean. A high temperature of 55° is less likely to happen in July, when 55° is farther away from the mean.

### Section 5.2

**5. Shipments.**

a)  Adding 4 ounces will affect only the median. The new median will be 68 +4 = 72 ounces, and the IQR will remain at 40 ounces.

b)  Changing the units will affect both the median and IQR. The median will be 72/16 = 4.5 pounds and the IQR will be 40/16 = 2.5 pounds.

**7. Men's shoe sizes.**

a)  The mean US shoe size will be affected by both the multiplication and subtraction. $\overline{USsize} = \overline{EuroSize} \times 0.7865 - 24 = 44.65 \times 0.7865 - 24 \approx 11.12$. The average US shoe size for the respondents is about 11.12.

b)  The standard deviation of US shoe sizes will only be affected by the multiplication. Adding or subtracting a constant from each value doesn't affect the spread. $\sigma_{US} = \sigma_{Euro} \times 0.7865 = 2.03 \times 0.7865 \approx 1.597$. The standard deviation of the respondents' shoe sizes, in US units, is about 1.597.

**Section 5.3**

9. **Guzzlers?**

   a) The Normal model for auto fuel economy is at the right.

   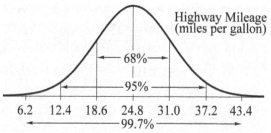

   b) Approximately 68% of the cars are expected to have highway fuel economy between 18.6 mpg and 31.0 mpg.

   c) Approximately 16% of the cars are expected to have highway fuel economy above 31 mpg.

   d) Approximately 13.5% of the cars are expected to have highway fuel economy between 31 mpg and 37 mpg.

   e) The worst 2.5% of cars are expected to have fuel economy below approximately 12.4 mpg.

11. **Checkup.**

    The boy's height is 1.88 standard deviations below the mean height of American children his age.

**Section 5.4**

13. **Normal cattle.**

    a)

    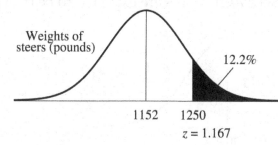

    $$z = \frac{y - \mu}{\sigma}$$

    $$z = \frac{1250 - 1152}{84}$$

    $$z \approx 1.167$$

    According to the Normal model, we expect 12.2% of steers to weigh over 1250 pounds.

    b)

    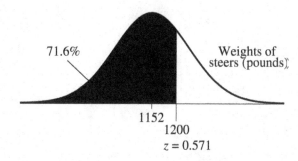

    $$z = \frac{y - \mu}{\sigma}$$

    $$z = \frac{1200 - 1152}{84}$$

    $$z \approx 0.571$$

    According to the Normal model, 71.6% of steers are expected to weigh under 1200 pounds.

c)

$$z = \frac{y - \mu}{\sigma}$$

$$z = \frac{1000 - 1152}{84}$$

$$z \approx -1.810$$

$$z = \frac{y - \mu}{\sigma}$$

$$z = \frac{1100 - 1152}{84}$$

$$z \approx -0.619$$

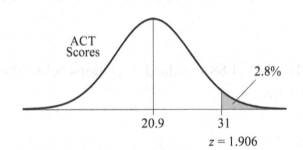

According to the Normal model, 23.3% of steers are expected to weigh between 1000 and 1100 pounds.

## 15. ACT scores.

a)

$$z = \frac{y - \mu}{\sigma}$$

$$z = \frac{31 - 20.9}{5.3}$$

$$z = 1.906$$

According to the Normal model, 2.8% of ACT scores are expected to be over 31.

b)

$$z = \frac{y - \mu}{\sigma}$$

$$z = \frac{18 - 20.9}{5.3}$$

$$z = -0.547$$

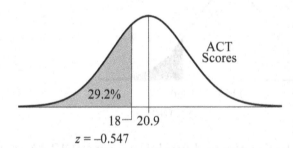

According to the Normal model, 29.2% of ACT scores are expected to be under 18.

**c)**

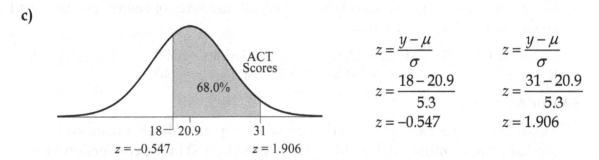

$$z = \frac{y - \mu}{\sigma}$$

$$z = \frac{18 - 20.9}{5.3}$$

$$z = -0.547$$

$$z = \frac{y - \mu}{\sigma}$$

$$z = \frac{31 - 20.9}{5.3}$$

$$z = 1.906$$

According to the Normal model, about 68.0% of ACT scores are between 18 and 31.

**Section 5.5**

**17. Music library.**

**a)** The Normal probability plot is not straight, so there is evidence that the distribution of the lengths of songs in Corey's music library is not Normal.

**b)** The distribution of the lengths of songs in Corey's music library appears to be skewed to the right. The Normal probability plot show that the longer songs in Corey's library are much longer than the lengths predicted by the Normal model. The song lengths are much longer than their quantile scores would predict for a Normal model.

**Chapter Exercises.**

**19. Payroll.**

**a)** The distribution of salaries in the company's weekly payroll is skewed to the right. The mean salary, $700, is higher than the median, $500.

**b)** The IQR, $600, measures the spread of the middle 50% of the distribution of salaries.

$$Q3 - Q1 = IQR$$
$$Q3 = Q1 + IQR$$
$$Q3 = \$350 + \$600$$
$$Q3 = \$950$$

50% of the salaries are found between $350 and $950.

**c)** If a $50 raise were given to each employee, all measures of center or position would increase by $50. The minimum would change to $350, the mean would change to $750, the median would change to $550, and the first quartile would change to $400. Measures of spread would not change. The entire distribution is simply shifted up $50. The range would remain at $1200, the IQR would remain at $600, and the standard deviation would remain at $400.

**d)** If a 10% raise were given to each employee, all measures of center, position, and spread would increase by 10%.

Minimum = $330   Mean = $770        Median = $550     Range = $1320
IQR = $660        First Quartile = $385   St. Dev. = $440

### 21. SAT or ACT?

Measures of center and position (lowest score, top 25% above, mean, and median) will be multiplied by 40 and increased by 150 in the conversion from ACT to SAT by the rule of thumb. Measures of spread (standard deviation and IQR) will only be affected by the multiplication.

Lowest score = 910        Mean = 1230       Standard deviation = 120
Top 25% above = 1350      Median = 1270     IQR = 240

### 23. Music library again.

*On the Nickel*, by Tom Waits has a *z*-score of 1.20

$$z = \frac{y - \mu}{\sigma}$$

$$z = \frac{380 - 242.4}{114.51}$$

$$z = 1.20$$

### 25. Combining test scores.

The *z*-scores, which account for the difference in the distributions of the two tests, are 1.5 and 0 for Derrick and 0.5 and 2 for Julie. Derrick's total is 1.5 which is less than Julie's 2.5.

### 27. Final Exams.

**a)** Anna's average is $\frac{83 + 83}{2} = 83$.   Megan's average is $\frac{77 + 95}{2} = 86$.

Only Megan qualifies for language honors, with an average higher than 85.

**b)** On the French exam, the mean was 81 and the standard deviation was 5. Anna's score of 83 was 2 points, or 0.4 standard deviations, above the mean. Megan's score of 77 was 4 points, or 0.8 standard deviations below the mean.

On the Spanish exam, the mean was 74 and the standard deviation was 15. Anna's score of 83 was 9 points, or 0.6 standard deviations, above the mean. Megan's score of 95 was 21 points, or 1.4 standard deviations, above the mean.

Measuring their performance in standard deviations is the only fair way in which to compare the performance of the two women on the test.

Anna scored 0.4 standard deviations above the mean in French and 0.6 standard deviations above the mean in Spanish, for a total of 1.0 standard deviation above the mean.

Megan scored 0.8 standard deviations below the mean in French and 1.4 standard deviations above the mean in Spanish, for a total of only 0.6 standard deviations above the mean.

Anna did better overall, but Megan had the higher average. This is because Megan did very well on the test with the higher standard deviation, where it was comparatively easy to do well.

**29. Cattle.**

a) A steer weighing 1000 pounds would be about 1.81 standard deviations below the mean weight. $z = \dfrac{y - \mu}{\sigma} = \dfrac{1000 - 1152}{84} \approx -1.81$

b) A steer weighing 1000 pounds is more unusual. Its $z$-score of –1.81 is further from 0 than the 1250 pound steer's $z$-score of 1.17.

**31. More cattle.**

a) The new mean would be 1152 – 1000 = 152 pounds. The standard deviation would not be affected by subtracting 1000 pounds from each weight. It would still be 84 pounds.

b) The mean selling price of the cattle would be 0.40(1152) = $460.80. The standard deviation of the selling prices would be 0.40(84) = $33.60.

**33. Cattle, part III.**

Generally, the minimum and the median would be affected by the multiplication and subtraction. The standard deviation and the IQR would only be affected by the multiplication.

Minimum = 0.40(980) – 20 = $372.00          Median = 0.40(1140) – 20 = $436
Standard deviation = 0.40(84) = $33.60          IQR = 0.40(102) = $40.80

**35. Professors.**

The standard deviation of the distribution of years of teaching experience for college professors must be 6 years. College professors can have between 0 and 40 (or possibly 50) years of experience. A workable standard deviation would cover most of that range of values with ±3 standard deviations around the mean. If the standard deviation were 6 months ($\frac{1}{2}$ year), some professors would have years of experience 10 or 20 standard deviations away from the mean, whatever it is. That isn't possible. If the standard deviation were 16 years, ±2 standard deviations would be a range of 64 years. That's way too high. The only reasonable choice is a standard deviation of 6 years in the distribution of years of experience.

**37. Small steer.**

Any weight more than 2 standard deviations below the mean, or less than 1152 – 2(84) = 984 pounds might be considered unusually low.  We would expect to see a steer below 1152 – 3(84) = 900 very rarely.

**39. Trees.**

a)  The Normal model for the distribution of tree diameters is at the right.

b)  Approximately 95% of the trees are expected to have diameters between 1.0 inch and 19.8 inches.

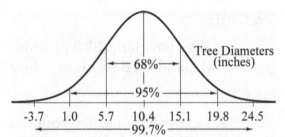

c)  Approximately 2.5% of the trees are expected to have diameters less than an inch.

d)  Approximately 34% of the trees are expected to have diameters between 5.7 inches and 10.4 inches.

e)  Approximately 16% of the trees are expected to have diameters over 15 inches.

**41. Trees, part II.**

The use of the Normal model requires a distribution that is unimodal and symmetric.  The distribution of tree diameters is neither unimodal nor symmetric, so use of the Normal model is not appropriate.

**43. Winter Olympics 2014.**

a)  The 2014 Winter Olympics downhill times have mean of 116.085 seconds and standard deviation 1.9215 seconds.  114.163 seconds is approximately 1 standard deviation below the mean.  If the Normal model is appropriate, 15.9% of the times should be below 114.163 seconds.

b)  4 out of 34 times (11.8%) are below 114.163 seconds.

c)  The percentages in parts a and b do not agree because the Normal model is not appropriate in this situation.

**d)** The histogram of 2014 Winter Olympic Downhill times is skewed to the right, with a cluster of high times as well. The Normal model is not appropriate for the distribution of times, because the distribution is not symmetric.

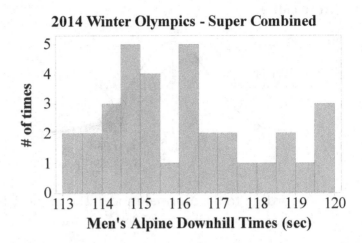

## 45. Receivers 2015.

**a)** Approximately 2.5% of the receivers are expected to gain more yards than 2 standard deviations above the mean number of yards gained.

**b)** The distribution of the number of yards gained has mean 274.73 yards and standard deviation 327.32 yards. According to the Normal model, we expect 2.5% of the receivers, or about 12 of them, to gain more than 2 standard deviations above the mean number of yards. This means more than 274.73 + 2(327.32) = 929.37 yards. In 2015, 31 receivers ran for more than 1122 yards.

**c)** The distribution of the number of yards run by wide receivers is skewed heavily to the right. Use of the Normal model is not appropriate for this distribution, since it is not symmetric.

## 47. CEO compensation sampled.

**a)** According to the Normal model, about 2.5% of CEOs would be expected to earn more than 2 standard deviation above the mean compensation.

**b)** The Normal model is not appropriate, since the distribution of CEO compensation is skewed to the right, not symmetric.

**c)** The Normal model is not appropriate, since the distribution of the sample means taken from samples of 30 CEO compensations is still skewed to the right, not symmetric.

**d)** The distribution of the sample means taken from samples of 100 CEO compensations is more symmetric, so the 68-95-99.7 Rule should work reasonably well.

**d)** The distribution of the sample means taken from samples of 200 CEO compensations is much more symmetric, so the 68-95-99.7 Rule should work quite well.

## 49. More cattle.

a)

$$z = \frac{y - \mu}{\sigma}$$

$$1.282 = \frac{y - 1152}{84}$$

$$y \approx 1259.7$$

Weights of steers (pounds)

10%

1152    $z = 1.282$
$y = 1259.7$ pounds

According to the Normal model, the highest 10% of steer weights are expected to be above approximately 1259.7 pounds.

b)

$$z = \frac{y - \mu}{\sigma}$$

$$-0.842 = \frac{y - 1152}{84}$$

$$y \approx 1081.3$$

20%

Weights of steers (pounds)

$z = -0.842$    1152
$y = 1081.3$ pounds

According to the Normal model, the lowest 20% of weights of steers are expected to be below approximately 1081.3 pounds.

c)

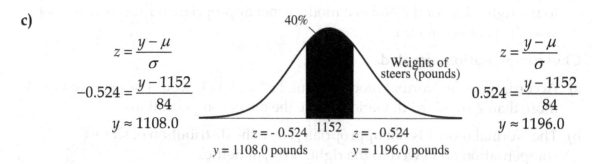

$$z = \frac{y - \mu}{\sigma}$$

$$-0.524 = \frac{y - 1152}{84}$$

$$y \approx 1108.0$$

40%

Weights of steers (pounds)

$$z = \frac{y - \mu}{\sigma}$$

$$0.524 = \frac{y - 1152}{84}$$

$$y \approx 1196.0$$

$z = -0.524$    1152    $z = -0.524$
$y = 1108.0$ pounds    $y = 1196.0$ pounds

According to the Normal model, the middle 40% of steer weights is expected to be between about 1108.0 pounds and 1196.0 pounds.

**51. Cattle, finis.**

a)

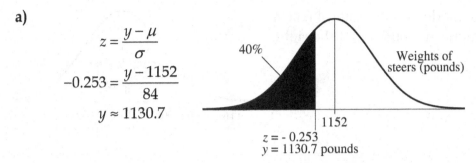

$$z = \frac{y - \mu}{\sigma}$$

$$-0.253 = \frac{y - 1152}{84}$$

$$y \approx 1130.7$$

*z* = - 0.253
*y* = 1130.7 pounds

According to the Normal model, the weight at the 40th percentile is 1130.7 pounds. This means that 40% of steers are expected to weigh less than 1130.7 pounds.

b)

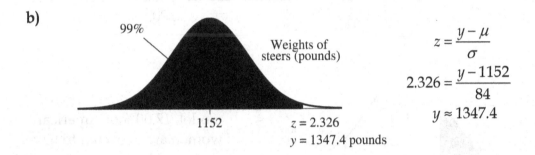

$$z = \frac{y - \mu}{\sigma}$$

$$2.326 = \frac{y - 1152}{84}$$

$$y \approx 1347.4$$

*z* = 2.326
*y* = 1347.4 pounds

According to the Normal model, the weight at the 99th percentile is 1347.4 pounds. This means that 99% of steers are expected to weigh less than 1347.4 pounds.

c)

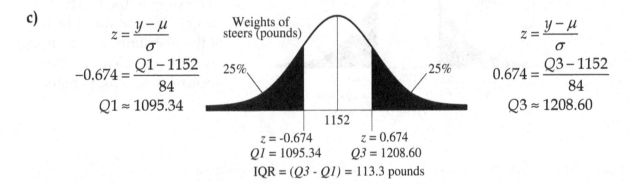

$$z = \frac{y - \mu}{\sigma}$$

$$-0.674 = \frac{Q1 - 1152}{84}$$

$$Q1 \approx 1095.34$$

$$z = \frac{y - \mu}{\sigma}$$

$$0.674 = \frac{Q3 - 1152}{84}$$

$$Q3 \approx 1208.60$$

*z* = -0.674      *z* = 0.674
*Q1* = 1095.34   *Q3* = 1208.60
IQR = (*Q3* - *Q1*) = 113.3 pounds

According to the Normal model, the IQR of the distribution of weights of Angus steers is about 113.3 pounds.

**53. Cholesterol.**

a)  The Normal model for cholesterol levels of adult American women is at the right.

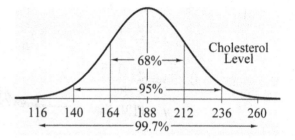

b)

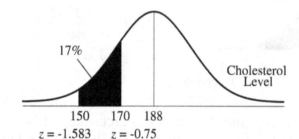

$$z = \frac{y - \mu}{\sigma}$$

$$z = \frac{200 - 188}{24}$$

$$z = 0.5$$

According to the Normal model, 30.85% of American women are expected to have cholesterol levels over 200.

c)

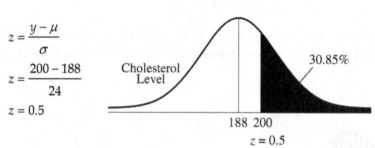

According to the Normal model, 17.00% of American women are expected to have cholesterol levels between 150 and 170.

d)

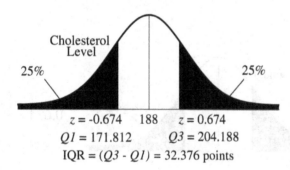

According to the Normal model, the interquartile range of the distribution of cholesterol levels of American women is approximately 32.38 points.

e)

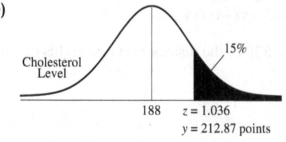

$$z = \frac{y - \mu}{\sigma}$$

$$1.036 = \frac{y - 188}{24}$$

$$y = 212.87$$

According to the Normal model, the highest 15% of women's cholesterol levels are above approximately 212.87.

## 55. Kindergarten.

**a)**

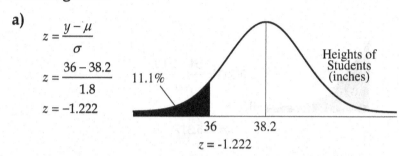

$$z = \frac{y - \mu}{\sigma}$$

$$z = \frac{36 - 38.2}{1.8}$$

$$z = -1.222$$

11.1%

Heights of Students (inches)

36        38.2

$z = -1.222$

According to the Normal model, approximately 11.1% of kindergarten kids are expected to be less than three feet (36 inches) tall.

**b)**

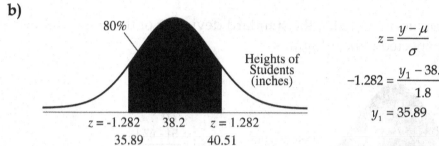

80%

Heights of Students (inches)

$z = -1.282$   38.2   $z = 1.282$
   35.89            40.51

$$z = \frac{y - \mu}{\sigma}$$

$$-1.282 = \frac{y_1 - 38.2}{1.8}$$

$$y_1 = 35.89$$

$$z = \frac{y - \mu}{\sigma}$$

$$1.282 = \frac{y_2 - 38.2}{1.8}$$

$$y_2 = 40.51$$

According to the Normal model, the middle 80% of kindergarten kids are expected to be between 35.89 and 40.51 inches tall. (The appropriate values of $z = \pm 1.282$ are found by using right and left tail percentages of 10% of the Normal model.)

**c)**

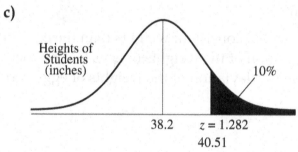

Heights of Students (inches)

38.2   $z = 1.282$
        40.51

10%

$$z = \frac{y - \mu}{\sigma}$$

$$1.282 = \frac{y - 38.2}{1.8}$$

$$y = 40.51$$

According to the Normal model, the tallest 10% of kindergarteners are expected to be at least 40.51 inches tall.

**57. Eggs.**

a)

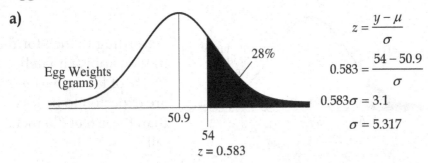

$$z = \frac{y - \mu}{\sigma}$$

$$0.583 = \frac{54 - 50.9}{\sigma}$$

$$0.583\sigma = 3.1$$

$$\sigma = 5.317$$

According to the Normal model, the standard deviation of the egg weights for young hens is expected to be 5.3 grams.

b)

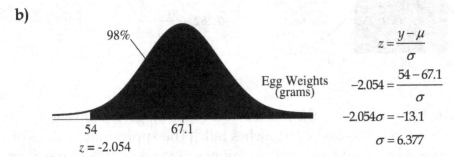

$$z = \frac{y - \mu}{\sigma}$$

$$-2.054 = \frac{54 - 67.1}{\sigma}$$

$$-2.054\sigma = -13.1$$

$$\sigma = 6.377$$

According to the Normal model, the standard deviation of the egg weights for older hens is expected to be 6.4 grams.

c) The younger hens lay eggs that have more consistent weights than the eggs laid by the older hens. The standard deviation of the weights of eggs laid by the younger hens is lower than the standard deviation of the weights of eggs laid by the older hens.

# Review of Part I – Exploring and Understanding Data

**R1.1. Bananas.**

a) A histogram of the prices of bananas from 15 markets, as reported by the USDA, appears at the right.

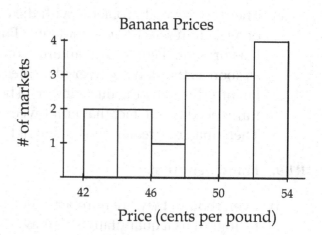

b) The distribution of banana prices is skewed to the left, so median and IQR are appropriate measures of center and spread. Median = 49 cents per pound
IQR = 6 cents per pound

c) The distribution of the prices of bananas from 15 markets, as reported by the USDA, is unimodal and skewed to the left. The center of the distribution is approximately 50 cents, with the lowest price 42 cents per pound and the highest price 53 cents per pound.

**R1.3. Singers by parts.**

a) The two statistics could be the same if there were many sopranos of that height.

b) The distribution of heights of each voice part is roughly symmetric. The basses and tenors are generally taller than the altos and sopranos, with the basses being slightly taller than the tenors. The sopranos and altos have about the same median height. Heights of basses and sopranos are more consistent than altos and tenors.

**R1.5. Beanstalks.**

a) The greater standard deviation for the distribution of women's heights means that their heights are more variable than the heights of men.

b) The z-score for women to qualify is 2.4 compared with 1.75 for men, so it is harder for women to qualify.

**R1.7. State University.**

a) *Who* – Local residents near State University. *What* – Age, whether or not the respondent attended college, and whether or not the respondent had a favorable opinion of State University. *When* – Not specified. *Where* – Region around State University. *Why* – The information will be included in a report to the University's directors. *How* – 850 local residents were surveyed by phone.

**b)** There is one quantitative variable, age, probably measured in years. There are two categorical variables, college attendance (yes or no), and opinion of State University (favorable or unfavorable).

**c)** There are several problems with the design of the survey. No mention is made of a random selection of residents. Furthermore, there may be a non-response bias present. People with an unfavorable opinion of the university may hang up as soon as the staff member identifies himself or herself. Also, response bias may be introduced by the interviewer. The responses of the residents may be influenced by the fact that employees of the university are asking the questions. There may be greater percentage of favorable responses than truly exist.

**R1.9. Fraud detection.**

**a)** Even though they are numbers, the NAICS code is a categorical variable. A histogram is a quantitative display, so it is not appropriate.

**b)** The Normal model will not work at all. The Normal model is for modeling distributions of unimodal and symmetric quantitative variables. NAICS code is a categorical variable.

**R1.11. Cramming.**

**a)** Comparitive boxplots of the distributions of Friday and Monday scores are at the right.

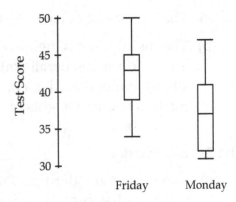

**b)** The distribution of scores on Friday was generally higher by about 5 points. Students fared worse on Monday after preparing for the test on Friday. The spreads are about the same, but the scores on Monday are slightly skewed to the right.

**c)** A histogram of the distribution of change in test score is at the right.

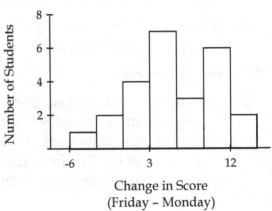

**d)** The distribution of changes in score is roughly unimodal and symmetric, and is centered near 4 points. Changes ranged from a student who scored 5 points higher on Monday, to two students who each scored 14 points higher on Friday. Only three students did better on Monday.

**R1.13. Let's play cards.**

a) Suit is a categorical variable.

b) In the game of Go Fish, the denomination is not ordered. Numbers are merely matched with one another. You may have seen children's Go Fish decks that have symbols or pictures on the cards instead of numbers. These work just fine.

c) In the game of Gin Rummy, the order of the cards is important. During the game, ordered "runs" of cards are assembled (with Jack = 11, Queen = 12, King = 13), and at the end of the hand, points are totaled from the denomination of the card (face cards = 10 points). However, even in Gin Rummy, the denomination of the card sometimes behaves like a categorical variable. When you are collecting 3s, for example, order doesn't matter.

**R1.15. Hard water.**

a) The variables in this study are both quantitative. Annual mortality rate for males is measured in deaths per 100,000. Calcium concentration is measured in parts per million.

b) The distribution of calcium concentration is skewed right, with many towns having concentrations below 25 ppm. The rest of the towns have calcium concentrations which are distributed in a fairly uniform pattern from 25 ppm to 100 ppm, tapering off to a maximum concentration around 150 ppm. The distribution of mortality rates is unimodal and symmetric, with center approximately 1500 deaths per 100,000. The distribution has a range of 1000 deaths per 100,000, from 1000 to 2000 deaths per 100,000.

**R1.17. Seasons.**

a) The two histograms have different horizontal and vertical scales. This makes a quick comparison impossible.

b) The center of the distribution of average temperatures in January in is the low 30s, compared to a center of the distribution of July temperatures in the low 70s. The January distribution is also much more spread out than the July distribution. The range is over 50 degrees in January, compared to a range of over 20 in July. The distribution of average temperature in January is skewed slightly to the right, while the distribution of average temperature in July is roughly symmetric.

c) The distribution of difference in average temperature (July – January) for 60 large U.S. cities is slightly skewed to the left, with median at approximately 44 degrees. There are several low outliers, cities with very little difference between their average July and January temperatures. The single high outlier is a city with a large difference in average temperature between July and January. The middle 50% of differences are between approximately 38 and 46 degrees.

### R1.19. Old Faithful?

a) The distribution of duration of the 222 eruptions is bimodal, with modes at approximately 2 minutes and 4.5 minutes. The distribution is fairly symmetric around each mode.

b) The bimodal shape of the distribution of duration of the 222 eruptions suggests that there may be two distinct groups of eruption durations. Summary statistics would try to summarize these two groups as a single group, which wouldn't make sense.

c) The intervals between eruptions are generally longer for long eruptions than the intervals for short eruptions. Over 75% of the short eruptions had intervals of approximately 60 minutes or less, while almost all of the long eruptions had intervals of more than 60 minutes.

### R1.21. Liberty's nose.

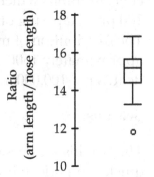

a) The distribution of the ratio of arm length to nose length of 18 girls in a statistics class is unimodal and roughly symmetric, with center around 15. There is one low outlier, a ratio of 11.8. A boxplot is provided at the right. A histogram or stemplot is also an appropriate display.

b) In the presence of an outlier, the 5-number summary is the appropriate choice for summary statistics. The 5-number summary is 11.8, 14.4, 15.25, 15.7, 16.9. The IQR is 1.3.

c) The ratio of 9.3 for the Statue of Liberty is very low, well below the lowest ratio in the statistics class, 11.8, which is already a low outlier. Compared to the girls in the statistics class, the Statue of Liberty's nose is very long in relation to her arm.

**R1.23. Sample.**

Overall, the follow-up group was insured only 11.1% of the time as compared to 16.6% for the not traced group. At first, it appears that group is associated with presence of health insurance. But for blacks, the follow-up group was quite close (actually slightly higher) in terms of being insured: 8.9% to 8.7%. The same is true for whites. The follow-up group was insured 83.3% of the time, compared to 82.5% of the not traced group. When broken down by race, we see that group is not associated with presence of health insurance for either race. This demonstrates Simpson's paradox, because the overall percentages lead us to believe that there is an association between health insurance and group, but we see the truth when we examine the situation more carefully.

**R1.25. Be quick!**

a) The Normal model for the distribution of reaction times is at the right.

b) The distribution of reaction times is unimodal and symmetric, with mean 1.5 seconds, and standard deviation 0.18 seconds. According to the Normal model, 95% of drivers are expected to have reaction times between 1.14 seconds and 1.86 seconds.

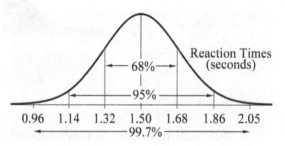

c)

$$z = \frac{y - \mu}{\sigma}$$

$$z = \frac{1.25 - 1.50}{0.18}$$

$$z = -1.389$$

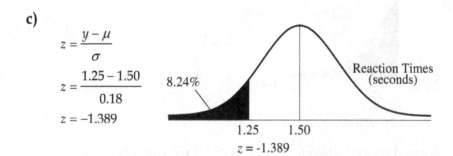

According to the Normal model, 8.24% of drivers are expected to have reaction times below 1.25 seconds.

d)

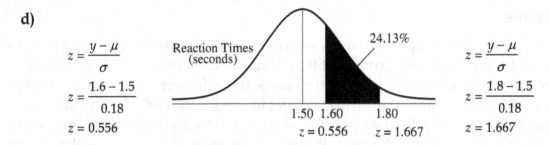

$$z = \frac{y - \mu}{\sigma}$$

$$z = \frac{1.6 - 1.5}{0.18}$$

$$z = 0.556$$

$$z = \frac{y - \mu}{\sigma}$$

$$z = \frac{1.8 - 1.5}{0.18}$$

$$z = 1.667$$

According to the Normal model, 24.13% of drivers are expected to have reaction times between 1.6 seconds and 1.8 seconds.

e)

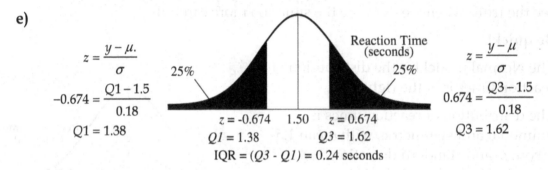

$$z = \frac{y - \mu.}{\sigma}$$

$$-0.674 = \frac{Q1 - 1.5}{0.18}$$

$$Q1 = 1.38$$

$$z = \frac{y - \mu}{\sigma}$$

$$0.674 = \frac{Q3 - 1.5}{0.18}$$

$$Q3 = 1.62$$

According to the Normal model, the interquartile range of the distribution of reaction times is expected to be 0.24 seconds.

f)

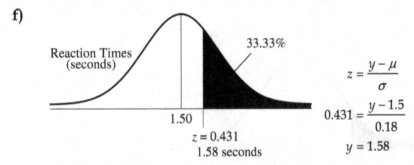

$$z = \frac{y - \mu}{\sigma}$$

$$0.431 = \frac{y - 1.5}{0.18}$$

$$y = 1.58$$

According to the Normal model, the slowest 1/3 of all drivers are expected to have reaction times of 1.58 seconds or more. (Remember that a high reaction time is a SLOW reaction time!)

**R1.27. Mail.**

a)  A histogram of the number of pieces of mail received at a school office is at the right.

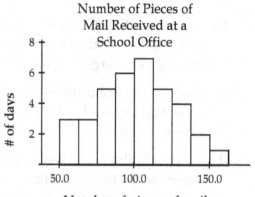

Number of Pieces of Mail Received at a School Office

b)  Since the distribution of number of pieces of mail is unimodal and symmetric, the mean and standard deviation are appropriate measures of center and spread. The mean number of pieces of mail is 100.25, and the standard deviation is 25.54 pieces.

c)  The distribution of the number of pieces of mail received at the school office is unimodal and symmetric, with mean 100.25 and standard deviation 25.54. The lowest number of pieces received in a day was 52 and the highest was 151.

d)  23 of the 36 days (64%) had a number of pieces of mail received within one standard deviation of the mean, or within the interval 74.71 - 125.79. This is fairly close to the 68% predicted by the Normal model. The Normal model may be useful for modeling the number of pieces received by this school office.

**R1.29. Herbal medicine.**

a)  *Who* – 100 customers. *What* – Researchers asked whether or not the customer had taken the cold remedy and had customers rate the effectiveness of the remedy on a scale from 1 to 10. *When* – Not specified. *Where* – Store where natural health products are sold. *Why* – The researchers were from the Herbal Medicine Council, which sounds suspiciously like a group that might be promoting the use of herbal remedies. *How* – Researchers conducted personal interviews with 100 customers. No mention was made of any type of random selection.

b)  "Have you taken the cold remedy?" is a categorical variable. Effectiveness on a scale of 1 to 10 is a categorical variable, as well, with respondents rating the remedy by placing it into one of 10 categories.

c)  Very little confidence can be placed in the Council's conclusions. Respondents were people who already shopped in a store that sold natural remedies. They may be pre-disposed to thinking that the remedy was effective. Furthermore, no attempt was made to randomly select respondents in a representative manner. Finally, the Herbal Medicine Council has an interest in the success of the remedy.

**R1.31. Engines.**

a) The count of cars is 38.

b) The mean displacement is higher than the median displacement, indicating a distribution of displacements that is skewed to the right. There are likely to be several very large engines in a group that consists of mainly smaller engines.

c) Since the distribution is skewed, the median and IQR are useful measures of center and spread. The median displacement is 148.5 cubic inches and the IQR is 126 cubic inches.

d) Your neighbor's car has an engine that is bigger than the median engine, but 227 cubic inches is smaller than the third quartile of 231, meaning that at least 25% of cars have a bigger engine than your neighbor's car. Don't be impressed!

e) Using the Outlier Rule (more than 1.5 IQRs beyond the quartiles) to find the fences:

Upper Fence: $Q3 + 1.5(IQR) = 231 + 1.5(126) = 420$ cubic inches.

Lower Fence: $Q1 - 1.5(IQR) = 105 - 1.5(126) = -84$ cubic inches.

Since there are certainly no engines with negative displacements, there are no low outliers. $Q1 + Range = 105 + 275 = 380$ cubic inches. This means that the maximum must be less than 380 cubic inches. Therefore, there are no high outliers (engines over 420 cubic inches).

f) It is not reasonable to expect 68% of the car engines to measure within one standard deviation of the mean. The distribution engine displacements is skewed to the right, so the Normal model is not appropriate.

g) Multiplying each of the engine displacements by 16.4 to convert cubic inches to cubic centimeters would affect measures of position and spread. All of the summary statistics (except the count!) could be converted to cubic centimeters by multiplying each by 16.4.

**R1.33. Age and party 2011.**

a) 3705 of 8414, or approximately 44.0%, of all voters surveyed were Republicans or leaned Republican.

b) This was a representative telephone survey conducted by Pew, a reputable polling firm. It is likely to be a reasonable estimate of the percentage of all voters who are Republicans.

c) $815 + 2416 = 3231$ of 8414, or approximately 38.4%, of all voters surveyed were under 30 or over 65 years old.

d) 73 of 8414, or approximately 0.87%, of all voters surveyed were classified as "Neither" and under the age of 30.

e) 73 of the 733 people classified as "Neither", or 9.96%, were under the age of 30.

**f)** 73 of the 815 respondents under 30, or 8.96%, were classified as "Neither".

### R1.35. Age and party 2011 II.

**a)** The marginal distribution of party affiliation is:
Republican – 44.0%  Democrat – 47.3%    Neither – 8.7%
(As counts:  Republican – 3705     Democrat – 3976     Independent – 733)

**b)** Graphs are at the right.

**c)** It appears that older voters are more likely to lean Republican, and younger voters are more likely to lean Deomocrat.

**d)** No. There is an evidence of an association between party affiliation and age. Younger voters tend to be more Democratic and less Republican.

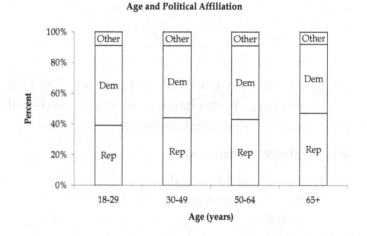

### R1.37. Some assembly required.

**a)**

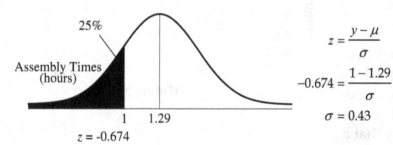

$$z = \frac{y - \mu}{\sigma}$$

$$-0.674 = \frac{1 - 1.29}{\sigma}$$

$$\sigma = 0.43$$

According to the Normal model, the standard deviation is 0.43 hours.

**b)**

$$z = \frac{y - \mu}{\sigma}$$

$$0.253 = \frac{y - 1.29}{0.43}$$

$$y = 1.40$$

According to the Normal model, the company would need to claim that the desk takes "less than 1.40 hours to assemble", not the catchiest of slogans!

**c)**

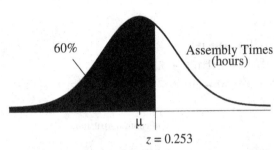

$$z = \frac{y - \mu}{\sigma}$$

$$0.253 = \frac{1 - \mu}{0.43}$$

$$\mu = 0.89$$

According to the Normal model, the company would have to lower the mean assembly time to 0.89 hour (53.4 minutes).

**d)** The new instructions and part-labeling may have helped lower the mean, but it also may have changed the standard deviation, making the assembly times more consistent as well as lower.

**R1.39 Shelves shuffled.**

**a)** The 21 middle shelf cereals have a mean sugar content of 9.62 g/serving, compared with 5.93 g/serving for the others, for a difference of 3.69 g/serving. The medians are 12 and 6 g/serving, respectively.

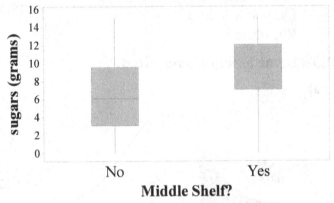

**b)** Answers will vary slightly. Because none of the 1000 shuffled differences were as large as the observed difference, we can say that a difference that large is unlikely to be produced by chance. There is evidence to suggest that the cereals on the middle shelf have a higher mean sugar content.

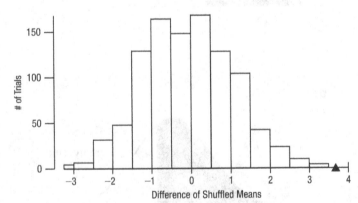

**R1.41. Hopkins Forest investigation.**

Answers will vary.

**R1.43 Student survey investigation.**

Answers will vary.

## Chapter 6 – Scatterplots, Association, and Correlation

### Section 6.1

**1. Association.**

**a)** Either weight in grams or weight in ounces could be the explanatory or response variable. Greater weights in grams correspond with greater weights in ounces. The association between weight of apples in grams and weight of apples in ounces would be positive, straight, and perfect. Each apple's weight would simply be measured in two different scales. The points would line up perfectly.

**b)** Circumference is the explanatory variable, and weight is the response variable, since one-dimensional circumference explains three-dimensional volume (and therefore weight). For apples of roughly the same size, the association would be positive, straight, and moderately strong. If the sample of apples contained very small and very large apples, the association's true curved form would become apparent.

**c)** There would be no association between shoe size and GPA of college freshmen.

**d)** Number of miles driven is the explanatory variable, and gallons remaining in the tank is the response variable. The greater the number of miles driven, the less gasoline there is in the tank. If a sample of different cars is used, the association is negative, straight, and moderate. If the data is gathered on different trips with the same car, the association would be strong.

**3. Bookstore sales.**

**a)** The scatterplot is to the right.

**b)** There is a positive association between bookstore sales and the number of sales people working.

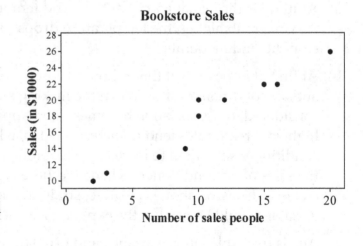

**c)** There is a linear association between bookstore sales and the number of sales people working.

**d)** There is a strong association between bookstore sales and the number of sales people working.

**e)** The relationship between bookstore sales and the number of sales people working has no outliers.

## Section 6.2

5. **Correlation facts.**

   a) True.

   b) False. The correlation will remain the same.

   c) False. Correlation has no units.

## Section 6.3

7. **Bookstore sales again.**

   This conclusion is not justified. Correlation does not demonstrate causation. The analyst argues that the number of sales staff working causes sales to be higher. It is possible (perhaps more plausible) that the store hired more people as sales increased. The causation may run in the opposite direction of the analyst's argument.

## Section 6.4

9. **Salaries and logs.**

   Since $\log_{10} 10,000 = 4$, $\log_{10} 100,000 = 5$, and $\log_{10} 1,000,000 = 6$, the plotted points will be (1, 4), (15, 5), and (30, 6). The plot of these three points will lie very close to a straight line.

## Chapter Exercises.

11. **Association III.**

    a) Altitude is the explanatory variable, and temperature is the response variable. As you climb higher, the temperature drops. The association is negative, straight, and moderate.

    b) At first, it appears that there should be no association between ice cream sales and air conditioner sales. When the lurking variable of temperature is considered, the association becomes more apparent. When the temperature is high, ice cream sales tend to increase. Also, when the temperature is high, air conditioner sales tend to increase. Therefore, there is likely to be an increase in the sales of air conditioners whenever there is an increase in the sales of ice cream. The association is positive, straight, and moderate. Either one of the variables could be used as the explanatory variable.

    c) Age is the explanatory variable, and grip strength is the response variable. The association is neither negative nor positive, but is curved, and moderate in strength, due to the variability in grip strength among people in general. The very young would have low grip strength, and grip strength would increase as age increased. After reaching a maximum (at whatever age physical prowess peaks), grip strength would decline again, with the elderly having low grip strengths.

d) Blood alcohol content is the explanatory variable, and reaction time is the response variable. As blood alcohol level increase, so does the time it takes to react to a stimulus. The association is positive, probably curved, and strong. The scatterplot would probably be almost linear for low concentrations of alcohol in the blood, and then begin to rise dramatically, with longer and longer reaction times for each incremental increase in blood alcohol content.

## 13. Scatterplots.

a) None of the scatterplots show little or no association, although # 4 is very weak.

b) #3 and #4 show negative association. Increases in one variable are generally related to decreases in the other variable.

c) #2, #3, and #4 each show a straight association.

d) #2 shows a moderately strong association.

e) #1 and #3 each show a very strong association. #1 shows a curved association and #3 shows a straight association.

## 15. Performance IQ scores vs. brain size.

The scatterplot of IQ scores *vs.* Brain Sizes is scattered, with no apparent pattern. There appears to be little or no association between the IQ scores and brain sizes displayed in this scatterplot.

## 17. Firing pottery.

a) A histogram of the number of broken pieces is at the right.

b) The distribution of the number broken pieces per batch of pottery is skewed right, centered around 1 broken piece per batch. Batches had from 0 and 6 broken pieces. The scatterplot does not show the center or skewness of the distribution.

Number of Broken Pieces per Batch (24 batches)

c) The scatterplot shows that the number of broken pieces increases as the batch number increases. If the 8 daily batches are numbered sequentially, this indicates that batches fired later in the day generally have more broken pieces. This information is not visible in the histogram.

## 19. Matching.

a) 0.006    b) 0.777    c) - 0.923    d) - 0.487

### 21. Politics.

The candidate might mean that there is an **association** between television watching and crime. The term correlation is reserved for describing linear associations between quantitative variables. We don't know what type of variables "television watching" and "crime" are, but they seem categorical. Even if the variables are quantitative (hours of TV watched per week, and number of crimes committed, for example), we aren't sure that the relationship is linear. The politician also seems to be implying a cause-and-effect relationship between television watching and crime. Association of any kind does not imply causation.

### 23. Coasters 2015.

a) It is appropriate to calculate correlation. Both height of the drop and speed are quantitative variables, the scatterplot shows an association that is straight enough, and there are not outliers.

b) There is a strong, positive, straight association between drop and speed; the greater the height of the initial drop, the higher the top speed.

### 25. Streams and hard water.

It is not appropriate to summarize the strength of the association between water hardness and pH with a correlation, since the association is curved, not Straight Enough.

### 27. Cold nights.

The correlation is between the number of days since January 1 and temperature is likely to be near zero. We expect the temperature to be low in January, increase through the spring and summer, then decrease again. The relationship is not Straight Enough, so correlation is not an appropriate measure of strength.

### 29. Prediction units.

The correlation between prediction error and year would not change, since the correlation is based on $z$-scores. The $z$-scores are the same whether the prediction errors are measured in nautical miles or miles.

### 31. Correlation errors.

a) If the association between GDP and infant mortality is linear, a correlation of – 0.772 shows a moderate, negative association. Generally, as GDP increases, infant mortality rate decreases.

b) Continent is a categorical variable. Correlation measures the strength of linear associations between quantitative variables.

## 33. Height and reading.

**a)** Actually, this *does* mean that taller children in elementary school are better readers. However, this does *not* mean that height causes good reading ability.

**b)** Older children are generally both taller and are better readers. Age is the lurking variable.

## 35. Correlations conclusions I.

**a)** No. We don't know this from correlation alone. The relationship between age and income may be non-linear, or the relationship may contain outliers.

**b)** No. We can't tell the form of the relationship between age and income. We need to look at the scatterplot.

**c)** No. The correlation between age and income doesn't tell us anything about outliers.

**d)** Yes. Correlation is based on *z*-scores, and is unaffected by changes in units.

## 37. Baldness and heart disease.

Even though the variables baldness and heart disease were assigned numerical values, they are categorical. Correlation is only an appropriate measure of the strength of linear association between quantitative variables. Their conclusion is meaningless.

## 39. Income and housing.

**a)** There is a positive, moderately strong, linear relationship between *Housing Cost Index* and *Median Family Income*, with several states whose *Housing Cost Index* seems high for their *Median Family Income*, and one state whose *Housing Cost Index* seems low for their *Median Family Income*.

**b)** Correlation is based on *z*-scores. The correlation would still be 0.65.

**c)** Correlation is based on *z*-scores, and is unaffected by changes in units. The correlation would still be 0.65.

**d)** Washington D.C. would be a moderately high outlier, with *Housing Cost Index* high for its *Median Family Income*. Since it doesn't fit the pattern, the correlation would decrease slightly if Washington D.C. were included.

**e)** No. We can only say that higher *Housing Cost Index* scores are associated with higher *Median Family Income*, but we don't know why. There may be other variables at work.

**41. Fuel economy 2016.**

a) A scatterplot of expected fuel economy vs. displacement is at the right.

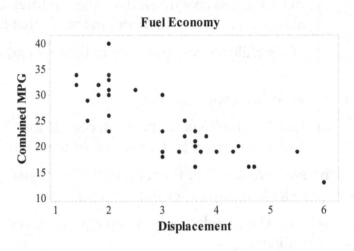

b) There is a strong, negative, straight association between displacement and mileage of the selected vehicles. There don't appear to be any outliers. All of the cars seem to fit the same pattern. Cars with larger engines tend to have lower mileage.

c) Since the relationship is linear, with no outliers, correlation is an appropriate measure of strength. The correlation between displacement and mileage of the selected vehicles is $r = -0.797$.

d) There is a strong linear relationship in the negative direction between displacement and highway gas mileage. Lower fuel efficiency is associated with larger engines.

**43. Burgers.**

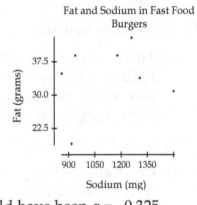

There is no apparent association between the number of grams of fat and the number of milligrams of sodium in several brands of fast food burgers. The correlation is only $r = 0.199$, which is close to zero, an indication of no association. One burger had a much lower fat content than the other burgers, at 19 grams of fat, with 920 milligrams of sodium. Without this (comparatively) low fat burger, the correlation would have been $r = -0.325$.

**45. Attendance 2016.**

a) Number of runs scored and attendance are quantitative variables, the relationship between them appears to be straight, and there are no outliers, so calculating a correlation is appropriate.

b) The association between attendance and runs scored is positive, straight, and moderate in strength. Generally, as the number of runs scored increases, so does attendance.

c) There is evidence of an association between attendance and runs scored, but a cause-and-effect relationship between the two is not implied. There may be lurking variables that can account for the increases in each. For example, perhaps winning teams score more runs and also have higher attendance. We don't have any basis to make a claim of causation.

**47. Coasters 2015 sampled.**

The distribution of the mean ride length of samples of 60 coasters is unimodal and symmetric, so we could use the 68–95–99.7 Rule for it. The distribution of median ride length of samples of 60 coasters is too skewed for the rule.

**49. Thrills 2013.**

The scatterplot at the right shows that the association between duration and length is straight, positive, and moderate, with no outliers. Generally, rides on coasters with a greater length tend to last longer. The correlation between length and duration is 0.736, indicating a moderate association.

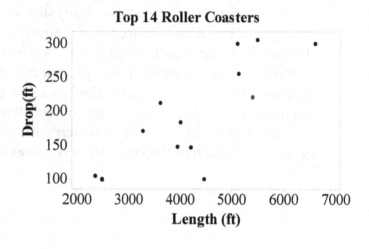

**51. Thrills III.**

a) Lower rank is better, so variables like speed and duration should have negative associations with rank. We would expect that as one variable (say length of ride) increases, the rank will improve, which means it will decrease.

b) Max vertical angle is the only variable with a negative correlation with rank, and that is due almost entirely to a single coaster, Nemesis.

### 53. Planets (more or less).

a) The association between Position Number of each planet and its distance from the sun (in millions of miles) is very strong, positive and curved. The scatterplot is at the right.

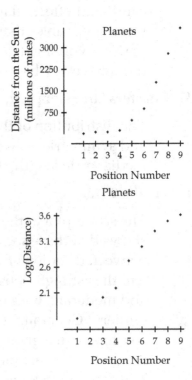

b) The relationship between Position Number and distance from the sun is not linear. Correlation is a measure of the degree of *linear* association between two variables.

c) The scatterplot of the logarithm of distance versus Position Number (shown at the right) still shows a strong, positive relationship, but it is straighter than the previous scatterplot. It still shows a curve in the scatterplot, but it is straight enough that correlation may now be used as an appropriate measure of the strength of the relationship between logarithm of distance and Position Number, which will in turn give an indication of the strength of the association.

## Chapter 7 – Linear Regression

### Section 7.1

**1. True or false.**

**a)** False. The line usually touches none of the points. The line minimizes the sum of least squares.

**b)** True.

**c)** False. Least squares means the sum of all the squared residuals is minimized.

### Section 7.2

**3. Least square interpretations.**

The weight of a newborn boy can be predicted as –5.94 kg plus 0.1875 kg per cm of length. This is a model fit to data. Parents should not be concerned if their newborn's length and weight don't fit this equation. No particular baby should be expected to fit this model exactly.

### Section 7.3

**5. Bookstore sales revisited.**

**a)** The slope of the line of best fit, $b_1$, is 0.914. (When using technology, the slope is 0.913. When calculated by hand, the standard deviations and correlation must be rounded, resulting in a slight inaccuracy.)

$$b_1 = r\frac{s_y}{s_x}$$
$$b_1 = (0.965)\frac{5.34}{5.64}$$
$$b_1 = 0.914$$

**b)** The model predicts an increase in sales of 0.914($1000), or $914, for each additional sales person working.

**c)** The intercept, $b_0$, is 8.09. (When using technology, the intercept is 8.10. When calculated by hand, the means and slope must be rounded, resulting in a slight inaccuracy.)

$$\hat{y} = b_0 + b_1 x$$
$$\bar{y} = b_0 + b_1 \bar{x}$$
$$17.6 = b_0 + (0.914)(10.4)$$
$$b_0 = 8.09$$

**d)** The model predicts that average sales would be approximately $8.09($1000), or $809, when there were no sales people working. This doesn't make sense in this context.

**e)** $\widehat{Sales} = 8.09 + 0.914\,People$ (hand calculation)

$\widehat{Sales} = 8.10 + 0.913\,People$ (technology)

**f)** $\widehat{Sales} = 8.09 + 0.914\,People = 8.09 + 0.914(18) = 24.542$

According to the model, we would expect sales to be approximately $24,540 when 18 sales people are working. ($24,530 using technology)

g) Residual $= sales - \widehat{sales} = \$25,000 - \$24,540 = \$460$ (Using technology, $470)

h) Since the residual is positive, we have underestimated the sales.

## Section 7.4

**7. Sophomore slump?**

The winners may be suffering from regression to the mean. Perhaps they weren't really better than other rookie executives, but just happened to have a lucky year the first year. When their second year performance landed them closer to the mean of the others, it looked like their performance had suffered.

## Section 7.5

**9. Bookstore sales once more.**

a) The residuals are measured in the same units as the response variable, thousands of dollars.

b) The residual with the largest magnitude, 2.77, contributes most to the sum of the squared residuals.

c) The residual with the smalles magnitude, 0.07, contributes least to the sum of the squared residuals.

## Section 7.6

**11. Bookstore sales last time.**

$R^2 = 93.12\%$ Approximately 93% of the variability in bookstore sales can be accounted for by the regression with the number of sales workers.

## Section 7.7

**13. Residual plots**

a) The residual plot has a clear curved pattern. The linearity assumption is violated.

b) One point on the residual plot has a much larger residual than the others. The outlier condition is violated.

c) The residual plot shows a fanned shape. The equal spread condition is violated.

## Chapter Exercises

**15. Cereals.**

$\widehat{Potassium} = 38 + 27\,Fiber = 38 + 27(9) = 281$ mg. According to the model, we expect cereal with 9 grams of fiber to have 281 milligrams of potassium.

## 17. More cereal.

A negative residual means that the potassium content is actually lower than the model predicts for a cereal with that much fiber.

## 19. Another bowl.

The model predicts that cereals will have approximately 27 more milligrams of potassium for each additional gram of fiber.

## 21. Cereal again.

$R^2 = (0.903)^2 \approx 0.815$   About 81.5% of the variability in potassium content is accounted for by the model.

## 23. Last bowl!

True potassium contents of cereals vary from the predicted values with a standard deviation of 30.77 milligrams.

## 25. Regression equations.

| $\bar{x}$ | $s_x$ | $\bar{y}$ | $s_y$ | $r$ | $\hat{y} = b_0 + b_1 x$ |
|---|---|---|---|---|---|
| a) 10 | 2 | 20 | 3 | 0.5 | $\hat{y} = 12.5 + 0.75x$ |
| b) 2 | 0.06 | 7.2 | 1.2 | -0.4 | $\hat{y} = 23.2 - 8x$ |
| c) 12 | 6 | 152 | 30 | -0.8 | $\hat{y} = 200 - 4x$ |
| d) 2.5 | 1.2 | 25 | 100 | 0.6 | $\hat{y} = -100 + 50x$ |

a)
$$b_1 = r\frac{s_y}{s_x}$$
$$b_1 = (0.5)\frac{3}{2}$$
$$b_1 = 0.75$$

$$\hat{y} = b_0 + b_1 x$$
$$\bar{y} = b_0 + b_1\bar{x}$$
$$20 = b_0 + 0.75(10)$$
$$b_0 = 12.5$$

b)
$$b_1 = r\frac{s_y}{s_x}$$
$$b_1 = (-0.4)\frac{1.2}{0.06}$$
$$b_1 = -8$$

$$\hat{y} = b_0 + b_1 x$$
$$\bar{y} = b_0 + b_1\bar{x}$$
$$7.2 = b_0 - 8(2)$$
$$b_0 = 23.2$$

c)
$$\hat{y} = b_0 + b_1 x$$
$$\bar{y} = b_0 + b_1\bar{x}$$
$$\bar{y} = 200 - 4(12)$$
$$\bar{y} = 152$$

$$b_1 = r\frac{s_y}{s_x}$$
$$-4 = (-0.8)\frac{s_y}{6}$$
$$s_y = 30$$

d)
$$\hat{y} = b_0 + b_1 x$$
$$\bar{y} = b_0 + b_1\bar{x}$$
$$\bar{y} = -100 + 50(2.5)$$
$$\bar{y} = 25$$

$$b_1 = r\frac{s_y}{s_x}$$
$$50 = r\frac{100}{1.2}$$
$$r = 0.6$$

## 27. Residuals.

a) The scattered residuals plot indicates an appropriate linear model.

b) The curved pattern in the residuals plot indicates that the linear model is not appropriate. The relationship is not linear.

c) The fanned pattern indicates that the linear model is not appropriate. The model's predicting power decreases as the values of the explanatory variable increase.

## 29. Real estate.

a) The explanatory variable ($x$) is size, measured in square feet, and the response variable ($y$) is price measured in thousands of dollars.

b) The units of the slope are thousands of dollars per square foot.

c) The slope of the regression line predicting price from size should be positive. Bigger homes are expected to cost more.

## 31. What slope?

The only slope that makes sense is 300 pounds per foot. 30 pounds per foot is too small. For example, a Honda Civic is about 14 feet long, and a Cadillac DeVille is about 17 feet long. If the slope of the regression line were 30 pounds per foot, the Cadillac would be predicted to outweigh the Civic by only 90 pounds! (The real difference is about 1500 pounds.) Similarly, 3 pounds per foot is too small. A slope of 3000 pounds per foot would predict a weight difference of 9000 pounds (4.5 tons) between Civic and DeVille. The only answer that is even reasonable is 300 pounds per foot, which predicts a difference of 900 pounds. This isn't very close to the actual difference of 1500 pounds, but at least it is in the right ballpark.

## 33. Real estate again.

71.4% of the variability in price can be accounted for by variability in size. (In other words, 71.4% of the variability in price can be accounted for by the linear model.)

## 35. Misinterpretations.

a) $R^2$ is an indication of the strength of the model, not the appropriateness of the model. A scattered residuals plot is the indicator of an appropriate model.

b) Regression models give predictions, not actual values. The student should have said, "The model predicts that a bird 10 inches tall is expected to have a wingspan of 17 inches."

## 37. Real estate redux.

a) The correlation between size and price is $r = \sqrt{R^2} = \sqrt{0.714} = 0.845$. The positive value of the square root is used, since the relationship is believed to be positive.

b) The price of a home that is one standard deviation above the mean size would be predicted to be 0.845 standard deviations (in other words $r$ standard deviations) above the mean price.

c) The price of a home that is two standard deviations below the mean size would be predicted to be 1.69 (or $2 \times 0.845$) standard deviations below the mean price.

## 39. ESP.

a) First, since no one has ESP, you must have scored 2 standard deviations above the mean by chance. On your next attempt, you are unlikely to duplicate the extraordinary event of scoring 2 standard deviations above the mean. You will likely "regress" towards the mean on your second try, getting a lower score. If you want to impress your friend, don't take the test again. Let your friend think you can read his mind!

b) Your friend doesn't have ESP, either. No one does. Your friend will likely "regress" towards the mean score on his second attempt, meaning his score will probably go up. If the goal is to get a higher score, your friend should try again.

## 41. More real estate.

a) According to the linear model, the price of a home is expected to increase $61 (0.061 thousand dollars) for each additional square-foot in size.

b) $\widehat{Price} = 47.82 + 0.061\,Size$

$\widehat{Price} = 47.82 + 0.061(3000)$

$\widehat{Price} = 230.82$

According to the linear model, a 3000 square-foot home is expected to have a price of $230,820.

c) $\widehat{Price} = 47.82 + 0.061\,Size$

$\widehat{Price} = 47.82 + 0.061(1200)$

$\widehat{Price} = 121.02$

According to the linear model, a 1200 square-foot home is expected to have a price of $121,020. The asking price is $121,020 - $6000 = $115,020. $6000 is the (negative) residual.

## 43. Cigarettes.

a) A linear model is probably appropriate. The residuals plot shows residuals that are larger than others, but there is no clear curvature.

b) 81.4% of the variability in nicotine level is accounted for by variability in tar content. (In other words, 81.4% of the variability in nicotine level is accounted for by the linear model.)

**45. Another cigarette.**

a) The correlation between tar and nicotine is $r = \sqrt{R^2} = \sqrt{0.814} = 0.902$. The positive value of the square root is used, since the relationship is believed to be positive. Evidence of the positive relationship is the positive coefficient of tar in the regression output.

b) The average nicotine content of cigarettes that are two standard deviations below the mean in tar content would be expected to be about $1.804$ ($2 \times 0.902$) standard deviations below the mean nicotine content.

c) Cigarettes that are one standard deviation above average in nicotine content are expected to be about 0.902 standard deviations (in other words, $r$ standard deviations) above the mean tar content.

**47. Last cigarette.**

a) $\widehat{Nicotine} = 0.148305 + 0.062163\,Tar$ is the equation of the regression line that predicts nicotine content from tar content of cigarettes.

b)

$\widehat{Nicotine} = 0.148305 + 0.062163\,Tar$

$\widehat{Nicotine} = 0.148305 + 0.062163(4)$

$\widehat{Nicotine} = 0.397$

The model predicts that a cigarette with 4 mg of tar will have about 0.397 mg of nicotine.

c) For each additional mg of tar, the model predicts an increase of 0.062 mg of nicotine.

d) The model predicts that a cigarette with no tar would have 0.148 mg of nicotine.

e)

$\widehat{Nicotine} = 0.148305 + 0.062163\,Tar$

$\widehat{Nicotine} = 0.148305 + 0.062163(7)$

$\widehat{Nicotine} = 0.583$

The model predicts that a cigarette with 7 mg of tar will have 0.583 mg of nicotine. If the residual is –0.05, the cigarette actually had 0.533 mg of nicotine.

**49. Income and housing revisited.**

a) Yes. Both housing cost index and median family income are quantitative. The scatterplot is Straight Enough, although there may be a few outliers. The spread increases a bit for states with large median incomes, but we can still fit a regression line.

**b)** Using the summary statistics given in the problem, calculate the slope and intercept:

$$b_1 = r\frac{s_{HCI}}{s_{MFI}}$$

$$b_1 = (0.624)\frac{119.07}{7003.55}$$

$$b_1 = 0.0106$$

$$\hat{y} = b_0 + b_1 x$$

$$\bar{y} = b_0 + b_1 \bar{x}$$

$$342.3 = b_0 + 0.0106(46210)$$

$$b_0 = -147.526$$

The regression equation that predicts HCI from MFI is

$$\widehat{HCI} = -147.526 + 0.0106 MFI$$

(from technology:

$$\widehat{HCI} = -148.15 + 0.0106 MFI )$$

**c)**

$$\widehat{HCI} = -147.526 + 0.0106 MFI$$

$$\widehat{HCI} = -147.526 + 0.0106(44993)$$

$$\widehat{HCI} = 329.40$$

The model predicts that a state with median family income of \$44993 have an average housing cost index of 329.40.

**d)** The prediction is 218.62 too low. Washington has a positive residual.

**e)** The correlation is the slope of the regression line that relates z-scores, so the regression equation would be $\hat{z}_{HCI} = 0.624 z_{MFI}$.

**f)** The correlation is the slope of the regression line that relates z-scores, so the regression equation would be $\hat{z}_{MFI} = 0.624 z_{HCI}$.

**51. Online clothes.**

**a)** Using the summary statistics given in the problem, calculate the slope and intercept:

$$b_1 = r\frac{s_{Total}}{s_{Age}}$$

$$b_1 = (0.037)\frac{253.62}{8.51}$$

$$b_1 = 1.1027$$

$$\hat{y} = b_0 + b_1 x$$

$$\bar{y} = b_0 + b_1 \bar{x}$$

$$572.52 = b_0 + 1.1027(29.67)$$

$$b_0 = 539.803$$

The regression equation that predicts total online clothing purchase amount from age is

$$\widehat{Total} = 539.803 + 1.103 Age$$

**b)** Yes. Both total purchases and age are quantitative variables, and the scatterplot is Straight Enough, even though it is quite flat. There are no outliers and the plot does not spread throughout the plot.

**c)**

$$\widehat{Total} = 539.803 + 1.103\,Age$$

$$\widehat{Total} = 539.803 + 1.103(18)$$

$$\widehat{Total} = 559.66$$

The model predicts that an 18 year old will have \$559.66 in total yearly online clothing purchases.

$$\widehat{Total} = 539.803 + 1.103\,Age$$

$$\widehat{Total} = 539.803 + 1.103(50)$$

$$\widehat{Total} = 594.95$$

The model predicts that a 50 year old will have \$594.95 in total yearly online clothing purchases.

**d)** $R^2 = (0.037)^2 \approx 0.0014 = 0.14\%$..

**e)** This model would not be useful to the company. The scatterplot is nearly flat. The model accounts for almost none of the variability in total yearly purchases.

**53. SAT scores.**

**a)** The association between SAT Math scores and SAT Verbal Scores was linear, moderate in strength, and positive. Students with high SAT Math scores typically had high SAT Verbal scores.

**b)** One student got a 500 Verbal and 800 Math. That set of scores doesn't seem to fit the pattern.

**c)** $r = 0.685$ indicates a moderate, positive association between SAT Math and SAT Verbal, but only because the scatterplot shows a linear relationship. Students who scored one standard deviation above the mean in SAT Math were expected to score 0.685 standard deviations above the mean in SAT Verbal. Additionally, $R^2 = (0.685)^2 = 0.469225$, so 46.9% of the variability in math score was accounted for by variability in verbal score.

**d)** The scatterplot of verbal and math scores shows a relationship that is straight enough, so a linear model is appropriate.

$$b_1 = r\,\frac{s_{Math}}{s_{Verbal}}$$

$$b_1 = (0.685)\frac{96.1}{99.5}$$

$$b_1 = 0.661593$$

$$\hat{y} = b_0 + b_1 x$$

$$\bar{y} = b_0 + b_1 \bar{x}$$

$$612.2 = b_0 + 0.661593(596.3)$$

$$b_0 = 217.692$$

The equation of the least squares regression line for predicting SAT Math score from SAT Verbal score is $\widehat{Math} = 217.692 + 0.662\,Verbal$.

e) For each additional point in verbal score, the model predicts an increase of 0.662 points in math score. A more meaningful interpretation might be scaled up. For each additional 10 points in verbal score, the model predicts an increase of 6.62 points in math score.

f)

$\widehat{Math} = 217.692 + 0.662\,Verbal$

$\widehat{Math} = 217.692 + 0.662(500)$

$\widehat{Math} = 548.692$

According to the model, a student with a verbal score of 500 was expected to have a math score of 548.692.

g)

$\widehat{Math} = 217.692 + 0.662\,Verbal$

$\widehat{Math} = 217.692 + 0.662(800)$

$\widehat{Math} = 747.292$

According to the model, a student with a verbal score of 800 was expected to have a math score of 747.292. She actually scored 800 on math, so her residual was 800 – 747.292 = 52.708 points

## 55. SAT, take 2.

a) $r = 0.685$. The correlation between SAT Math and SAT Verbal is a unitless measure of the degree of linear association between the two variables. It doesn't depend on the order in which you are making predictions.

b) The scatterplot of verbal and math scores shows a relationship that is straight enough, so a linear model is appropriate.

$b_1 = r\dfrac{s_{Verbal}}{s_{Math}}$

$b_1 = (0.685)\dfrac{99.5}{96.1}$

$b_1 = 0.709235$

$\hat{y} = b_0 + b_1 x$

$\bar{y} = b_0 + b_1\bar{x}$

$596.3 = b_0 + 0.709235(612.2)$

$b_0 = 162.106$

The equation of the least squares regression line for predicting SAT Verbal score from SAT Math score is: $\widehat{Verbal} = 162.106 + 0.709\,Math$

c) A positive residual means that the student's actual verbal score was higher than the score the model predicted for someone with the same math score.

d)

$\widehat{Verbal} = 162.106 + 0.709\,Math$

$\widehat{Verbal} = 162.106 + 0.709(500)$

$\widehat{Verbal} = 516.606$

According to the model, a person with a math score of 500 was expected to have a verbal score of 516.606 points.

**e)**

$$\widehat{Math} = 217.692 + 0.662\,Verbal$$

$$\widehat{Math} = 217.692 + 0.662\,(516.606)$$

$$\widehat{Math} = 559.685$$

According to the model, a person with a verbal score of 516.606 was expected to have a math score of 559.685 points.

**f)** The prediction in part e) does not cycle back to 500 points because the regression equation used to predict math from verbal is a different equation than the regression equation used to predict verbal from math. One was generated by minimizing squared residuals in the verbal direction, the other was generated by minimizing squared residuals in the math direction. If a math score is one standard deviation above the mean, its predicted verbal score regresses toward the mean. The same is true for a verbal score used to predict a math score.

**57. Wildfires 2015.**

**a)** The scatterplot shows a roughly linear relationship between the year and the number of wildfires, so the linear model is appropriate. The relationship is very weak, however.

**b)** The model predicts a decrease of an average of about 220 wildfires per year.

**c)** It seems reasonable to interpret the intercept. The model predicts about 78,792 wildfires in 1985, which is within the scope of the data, although it isn't very useful since we know the actual number of wildfires in 1985. There isn't much need for a prediction.

**d)** The standard deviation of the residuals is 12,397 fires. That's a large residual, considering that these years show between 60,000 and 90,000 fires per year. The association just isn't very strong.

**e)** The model only accounts for about 2.7% of the variability in the number of fires each year. The rest of the variability is due to other factors that we don't have data about.

**59. Used cars 2014.**

**a)** We are attempting to predict the price in dollars of used Toyota Corollas from their age in years. A scatterplot of the relationship is at the right.

**b)** There is a strong, negative, linear association between price and age of used Toyota Corollas.

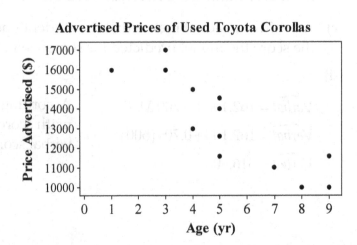

**Advertised Prices of Used Toyota Corollas**

**c)** The scatterplot provides evidence that the relationship is straight enough. A linear model will likely be an appropriate model.

**d)** Since $R^2 = 0.891$, simply take the square root to find $r$. $\sqrt{0.752} = 0.867$. Since association between age and price is negative, $r = -0.867$.

**e)** 75.2% of the variability in price of a used Toyota Corolla can be accounted for by variability in the age of the car.

**f)** The relationship is not perfect. Other factors, such as options, condition, and mileage explain the rest of the variability in price.

**61. More used cars 2014.**

**a)** The scatterplot from the previous exercise shows that the relationship is straight, so the linear model is appropriate.

```
Predictor      Coef  SE Coef      T      P
Constant    17674.0    836.2  21.14  0.000
Age          -844.5    146.1  -5.78  0.000


S = 1224.82    R-Sq = 75.2%    R-Sq(adj) = 73.0%
```

The regression equation to predict the price of a used Toyota Corolla from its age is

$\widehat{Price} = 17674 - 844.5\,Years$.

Computer regression output used is at the right.

**b)** According to the model, for each additional year in age, the car is expected to drop $844.5 in price.

**c)** The model predicts that a new Toyota Corolla (0 years old) will cost $17,674.

**d)** $\widehat{Price} = 17674 - 844.5\,Years$

$\widehat{Price} = 17674 - 844.5(7)$

$\hat{Price} = 11762.5$

According to the model, an appropriate price for a 7-year old Toyota Corolla is $11,762.50.

**e)** Buy the car with the negative residual. Its actual price is lower than predicted.

**f)** $\widehat{Price} = 17674 - 844.5\,Years$

$\widehat{Price} = 17674 - 844.5(10)$

$\widehat{Price} = 9229$

According to the model, a 10-year-old Corolla is expected to cost $9229. The car has an actual price of $8500, so its residual is $8500 − $9229 = − $729

**g)** The model would not be useful for predicting the price of a 25-year-old Corolla. The oldest car in the list is 9 years old. Predicting a price after 25 years would be an extrapolation.

### 63. Burgers revisited.

**a)** The scatterplot of calories vs. fat content in fast food hamburgers is at the right. The relationship appears linear, so a linear model is appropriate.

**Fat and Calories of Fast Food Burgers**

Dependent variable is:  Calories
No Selector
R squared = 92.3%   R squared (adjusted) = 90.7%
s = 27.33 with 7 - 2 = 5 degrees of freedom

| Source | Sum of Squares | df | Mean Square | F-ratio |
|---|---|---|---|---|
| Regression | 44664.3 | 1 | 44664.3 | 59.8 |
| Residual | 3735.73 | 5 | 747.146 | |

| Variable | Coefficient | s.e. of Coeff | t-ratio | prob |
|---|---|---|---|---|
| Constant | 210.954 | 50.10 | 4.21 | 0.0084 |
| Fat | 11.0555 | 1.430 | 7.73 | 0.0006 |

**b)** From the computer regression output, $R^2 = 92.3\%$. 92.3% of the variability in the number of calories can be explained by the variability in the number of grams of fat in a fast food burger.

**c)** From the computer regression output, the regression equation that predicts the number of calories in a fast food burger from its fat content is:

$$\widehat{Calories} = 210.954 + 11.0555\,Fat$$

**d)** The residuals plot at the right shows no pattern. The linear model appears to be appropriate.

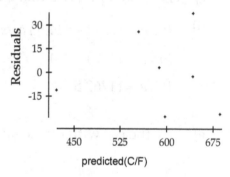

**e)** The model predicts that a fat free burger would have 210.954 calories. Since there are no data values close to 0, this extrapolation isn't of much use.

**f)** For each additional gram of fat in a burger, the model predicts an increase of 11.056 calories.

**g)** $\widehat{Calories} = 210.954 + 11.0555\,Fat = 210.954 + 11.0555(28) = 520.508$

The model predicts a burger with 28 grams of fat will have 520.508 calories. If the residual is +33, the actual number of calories is 520.508 + 33 ≈ 553.5 calories.

**65. A second helping of burgers.**

a) The model from Exercise 63 was for predicting number of calories from number of grams of fat. In order to predict grams of fat from the number of calories, a new linear model needs to be generated.

b) The scatterplot at the right shows the relationship between number fat grams and number of calories in a set of fast food burgers. The association is strong, positive, and linear. Burgers with higher numbers of calories typically have higher fat contents. The relationship is straight enough to apply a linear model.

Dependent variable is:    Fat
No Selector
R squared = 92.3%    R squared (adjusted) = 90.7%
s = 2.375 with 7 - 2 = 5 degrees of freedom

| Source | Sum of Squares | df | Mean Square | F-ratio |
|--------|----------------|-----|-------------|---------|
| Regression | 337.223 | 1 | 337.223 | 59.8 |
| Residual | 28.2054 | 5 | 5.64109 | |

| Variable | Coefficient | s.e. of Coeff | t-ratio | prob |
|----------|-------------|---------------|---------|------|
| Constant | -14.9622 | 6.433 | -2.33 | 0.0675 |
| Calories | 0.083471 | 0.0108 | 7.73 | 0.0006 |

The linear model for predicting fat from calories is:

$\widehat{Fat} = -14.9622 + 0.083471\,Calories$

The model predicts that for every additional 100 calories, the fat content is expected to increase by about 8.3 grams.

The residuals plot shows no pattern, so the model is appropriate. $R^2$ = 92.3%, so 92.3% of the variability in fat content can be accounted for by the model.

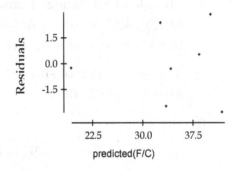

$\widehat{Fat} = -14.9622 + 0.083471\,Calories$

$\widehat{Fat} = -14.9622 + 0.083471\,(600)$

$\widehat{Fat} \approx 35.1$

According to the model, a burger with 600 calories is expected to have 35.1 grams of fat.

**67. New York bridges 2016.**

a) Overall, the model predicts the condition score of new bridges to be 5.0112, close to the cutoff of 5. The negative slope means that most bridges are predicted to have condition less than 5. The model is not a very encouraging one in regards to the conditions of New York City bridges.

b) According to the model, the condition of the bridges in New York City is decreasing by an average of 0.00513 per year. This is less rapid than the bridges in Tompkins County.

c) We shouldn't place too much faith in the model. $R^2$ of 3.9% is very low, and the standard deviation of the residuals, 0.6912, is quite high in relation to the scope of the data values themselves. This association is very weak.

**69. Climate change 2016.**

a) The correlation between $CO_2$ level and mean global temperature anomaly is $r = \sqrt{R^2} = \sqrt{0.897} = 0.9471$.

b) 89.7% of the variability in mean global temperature anomaly can be accounted for by variability in $CO_2$ level.

c) Since the scatterplot of $CO_2$ level and mean global temperature anomaly shows a relationship that is straight enough, use of the linear model is appropriate. The linear regression model that predicts mean global temperature anomaly from $CO_2$ level is: $\overline{TempAnomaly} = -3.17933 + 0.0099179\,CO_2$

d) The model predicts that an increase in $CO_2$ level of 1 ppm is associated with an increase of 0.0099°C in mean global temperature anomaly.

e) According to the model, the mean global temperature anomaly is predicted to be –3.179 °C when there is no $CO_2$ in the atmosphere. This is an extrapolation outside of the range of data, and isn't very meaningful in context, since there is always $CO_2$ in the atmosphere. We want to use this model to study the change in $CO_2$ level and how it relates to the change in temperature.

f) The residuals plot shows no apparent patterns. The linear model appears to be an appropriate one.

g)

$\overline{TempAnomaly} = -3.17933 + 0.0099179\,CO_2$

$\overline{TempAnomaly} = -3.17933 + 0.0099179(450)$

$\overline{TempAnomaly} = 1.283725$

According to the model, the mean global temperature anomaly is predicted to be 1.28 °C when the $CO_2$ level is 450 ppm.

**h)** No, this does not mean that when the $CO_2$ level hits 450 ppm, the temperature anomaly will be 1.28 °C. First, this is an extrapolation. The model can't say what will happen under other circumstances. Secondly, even for predictions within our set of data, the model can only give a general idea of what to expect.

**71. Body fat.**

**a)** The scatterplot of % body fat and weight of 20 male subjects, at the right, shows a strong, positive, linear association. Generally, as a subject's weight increases, so does % body fat. The association is straight enough to justify the use of the linear model.

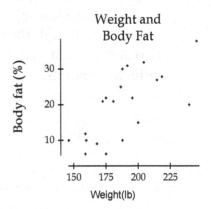

The linear model that predicts % body fat from weight is: $\widehat{\%Fat} = -27.3763 + 0.249874\,Weight$

**b)** The residuals plot, at the right, shows no apparent pattern. The linear model is appropriate.

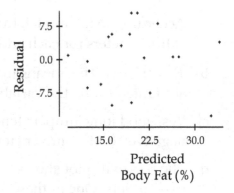

**c)** According to the model, for each additional pound of weight, body fat is expected to increase by about 0.25%.

**d)** Only 48.5% of the variability in % body fat can be accounted for by the model. The model is not expected to make predictions that are accurate.

**e)**

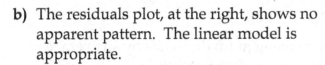

$\widehat{\%Fat} = -27.3763 + 0.249874\,Weight$

$\widehat{\%Fat} = -27.3763 + 0.249874(190)$

$\widehat{\%Fat} = 20.09976$

According to the model, the predicted body fat for a 190-pound man is 20.09976%.
The residual is $21 - 20.09976 \approx 0.9\%$.

## 73. Women's heptathlon revisited.

a) Both long jump height and 800 meter time are quantitative variables, the association is straight enough to use linear regression, though there are two outliers that may be influential points.

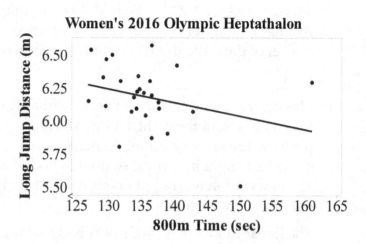

The regression equation to predict long jump from 800m results is:
$$\widehat{Longjump} = 7.59533 - 0.01033\,Time.$$

According to the model, the predicted long jump decreases by an average of 0.01033 meters for each additional second in 800 meter time.

b) $R^2 = 9.39\%$. This means that 9.39% of the variability in long jump distance is accounted for by the variability in 800 meter time.

c) Yes, good long jumpers tend to be fast runners. The slope of the association is negative. Faster runners tend to jump long distances, as well.

d) The residuals plot shows two outliers. One of these athletes, Akela Jones, had a long jump that was much longer than her 800m time would predict, making this point particularly influential. This model is not appropriate.

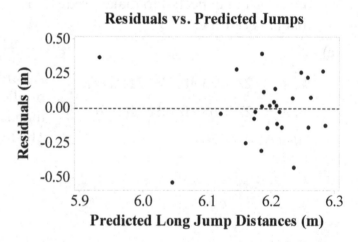

e) The linear model is not particularly useful for predicting long jump performance. First of all, there are outliers and influential points. Additionally, only 9.39% of the variability in long jump distance is accounted for by the variability in 800 meter time, leaving 90.61% of the variability accounted for by other variables. Finally, the residual standard deviation is 0.23 meters, which is not much smaller than the standard deviation of all long jumps, 0.2475 meters. Predictions are not likely to be accurate.

**75. Hard water.**

**a)** There is a fairly strong, negative, linear relationship between calcium concentration (in ppm) in the water and mortality rate (in deaths per 100,000). Towns with higher calcium concentrations tended to have lower mortality rates.

**b)** The linear regression model that predicts mortality rate from calcium concentration is $\widehat{Mortality} = 1676 - 3.23\,Calcium$.

**c)** The model predicts a decrease of 3.23 deaths per 100,000 for each additional ppm of calcium in the water. For towns with no calcium in the water, the model predicts a mortality rate of 1676 deaths per 100,000 people.

**d)** Exeter had 348.6 fewer deaths per 100,000 people than the model predicts.

**e)**

$\widehat{Mortality} = 1676 - 3.23\,Calcium$

$\widehat{Mortality} = 1676 - 3.23(100)$

$\widehat{Mortality} = 1353$

The town of Derby is predicted to have a mortality rate of 1353 deaths per 100,000 people.

**f)** 43% of the variability in mortality rate can be explained by variability in calcium concentration.

**77. Least squares.**

If the 4 $x$-values are plugged into $\hat{y} = 7 + 1.1x$, the 4 predicted values are $\hat{y} = 18, 29, 51$ and 62, respectively. The 4 residuals are –8, 21, –31, and 18. The squared residuals are 64, 441, 961, and 324, respectively. The sum of the squared residuals is 1790. Least squares means that no other line has a sum lower than 1790. In other words, it's the best fit.

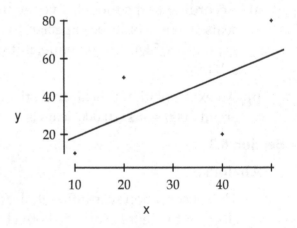

**79. Fuel Economy 2016 revisited.**

**a)** The distribution of the slopes is unimodal and slightly skewed to the low end. 95% of the random slopes fell between -4.4 and -2.2.

**b)** The slope for the sample from Chapter 6 is just barely inside the interval that holds the middle 95% of the randomly generated slopes, so by that definition, it is not unusual.

**c)** Answers will vary.

# Chapter 8 – Regression Wisdom

**Section 8.1**

**1. Credit card spending.**

The different segments are not scattered at random throughout the residual plot. Each segment may have a different relationship, which would affect the accuracy of any predictions made with the model that ignores the differences between segments.

**3. Market segments.**

Yes, it is clear that the relationship between January and December spending is not the same for all five segments. Using one overall model to predict January spending would be very misleading.

**Section 8.2**

**5. Cell phone costs.**

Your friend is extrapolating. It is impossible to know if a trend like this will continue so far into the future.

**7. Revenue and large venues.**

a)  According to the model, a venue that seats 10,000 would be expected to generate $354,472 in revenue, if it were to sell out.

$$\widehat{Revenue} = -14,228 + 36.87\,TicketSales$$
$$\widehat{Revenue} = -14,228 + 36.87(10,000)$$
$$\widehat{Revenue} = 354,472$$

b)  An extrapolation this far from the data is unreliable. We only have data up to about 3000 seats. 10,000 seats is well above that.

**Section 8.3**

**9. Abalone.**

This observation was influential. After it was removed, the correlation and the slope of the regression line both changed by a large amount.

**Section 8.4**

**11. Skinned knees.**

No. There is a lurking variable, seasonal temperature. In warm weather, more children will go outside and play, and if there are more children playing, there will be more skinned knees.

**Section 8.5**

**13. Grading.**

Individual student scores will vary greatly. The class averages will have much less variability and may disguise important patterns.

**Section 8.6**

**15. Residuals.**

a) The residuals plot shows no pattern. No re-expression is needed.

b) The residuals plot shows a curved pattern. Re-express to straighten the relationship.

c) The residuals plot shows a fan shape. Re-express to equalize spread.

**17. BK protein.**

The goal of this re-expression is to improve homoscedasticity. We desire more equal spread between groups.

**Section 8.7**

**19. BK protein again.**

The log re-expression is still preferable. The square root doesn't make the spreads as nearly equal. The reciprocal clearly goes too far on the Ladder of Powers.

**Chapter Exercises.**

**21. Marriage age 2015.**

a) The trend in age at first marriage for American women is very strong over the entire time period recorded on the graph, but the direction and form are different for different time periods. The trend appears to be somewhat linear, and consistent at around 22 years, up until about 1940, when the age seemed to drop dramatically, to under 21. From 1940 to about 1970, the trend appears non-linear and slightly positive. From 1975 to the present, the trend again appears linear and positive. The marriage age rose rapidly during this time period.

b) The association between age at first marriage for American women and year is strong over the entire time period recorded on the graph, but some time periods have stronger trends than others.

c) The correlation, or the measure of the degree of linear association is not high for this trend. The graph, as a whole, is non-linear. However, certain time periods, like 1975 to present, have a high correlation.

    **d)** Overall, the linear model is not appropriate. The scatterplot is not Straight Enough to satisfy the condition. You could fit a linear model to the time period from 1975 to 2003, but this seems unnecessary. The ages for each year are reported, and, given the fluctuations in the past, extrapolation seems risky.

### 23. Human Development Index 2015.

    **a)** Fitting a linear model to the association between HDI and GDPPC would be misleading, since the relationship is not straight.

    **b)** If you fit a linear model to these data, the residuals plot will be curved downward.

### 25. Good model?

    **a)** The student's reasoning is not correct. A scattered residuals plot, not high $R^2$, is the indicator of an appropriate model. Once the model is deemed appropriate, $R^2$ is used as a measure of the strength of the model.

    **b)** The model may not allow the student to make accurate predictions. The data may be curved, in which case the linear model would not fit well.

### 27. Movie dramas.

    **a)** The units for the slopes of these lines are millions of dollars per minutes of running time.

    **b)** The slopes of the regression lines are about the same. Dramas and movies from other genres have costs for longer movies that increase at the same rate.

    **c)** The regression line for dramas has a lower $y$-intercept. Regardless of running time, dramas cost about 20 million dollars less than other genres of movies of the same running time.

### 29. Oakland passengers 2016.

**a)** There are several features to comment on in this plot. There is a strong monthly pattern around the general trend. From 1997 to 2008, passengers increased fairly steadily with a notable exception of Sept. 2001, probably due to the attack on the twin towers. Then sometime in late 2008, departures dropped dramatically, possibly due to the economic crisis. Recently, they have been recovering, but not at the same rate as their previous increase.

**b)** The trend was fairly linear until late 2008, then passengers dropped suddenly.

**c)** The trend since 2009 has been linear (overall, ignoring monthly oscillations) If the increase continues to be linear, the predictions should be reasonable for the short term.

**31. Unusual points.**

a)  **1)**  The point has high leverage and a small residual.
  **2)**  The point is not influential. It has the *potential* to be influential, because its position far from the mean of the explanatory variable gives it high leverage. However, the point is not *exerting* much influence, because it reinforces the association.

  **3)**  If the point were removed, the correlation would become weaker. The point heavily reinforces the positive association. Removing it would weaken the association.
  **4)**  The slope would remain roughly the same, since the point is not influential.

b)  **1)**  The point has high leverage and probably has a small residual.
  **2)**  The point is influential. The point alone gives the scatterplot the appearance of an overall negative direction, when the points are actually fairly scattered.
  **3)**  If the point were removed, the correlation would become weaker. Without the point, there would be very little evidence of linear association.
  **4)**  The slope would increase, from a negative slope to a slope near 0. Without the point, the slope of the regression line would be nearly flat.

c)  **1)**  The point has moderate leverage and a large residual.
  **2)**  The point is somewhat influential. It is well away from the mean of the explanatory variable, and has enough leverage to change the slope of the regression line, but only slightly.
  **3)**  If the point were removed, the correlation would become stronger. Without the point, the positive association would be reinforced.
  **4)**  The slope would increase slightly, becoming steeper after the removal of the point. The regression line would follow the general cloud of points more closely.

d)  **1)**  The point has little leverage and a large residual.
  **2)**  The point is not influential. It is very close to the mean of the explanatory variable, and the regression line is anchored at the point $(\bar{x}, \bar{y})$, and would only pivot if it were possible to minimize the sum of the squared residuals. No amount of pivoting will reduce the residual for the stray point, so the slope would not change.
  **3)**  If the point were removed, the correlation would become slightly stronger, decreasing to become more negative. The point detracts from the overall pattern, and its removal would reinforce the association.
  **4)**  The slope would remain roughly the same. Since the point is not influential, its removal would not affect the slope.

### 33. The extra point.

1) Point e is very influential. Its addition will give the appearance of a strong, negative correlation like $r = -0.90$.

2) Point d is influential (but not as influential as point e). Its addition will give the appearance of a weaker, negative correlation like $r = -0.40$.

3) Point c is directly below the middle of the group of points. Its position is directly below the mean of the explanatory variable. It has no influence. Its addition will leave the correlation the same, $r = 0.00$.

4) Point b is almost in the center of the group of points, but not quite. Its addition will give the appearance of a very slight positive correlation like $r = 0.05$.

5) Point a is very influential. Its addition will give the appearance of a strong, positive correlation like $r = 0.75$.

### 35. What's the cause?

1) High blood pressure may cause high body fat.

2) High body fat may cause high blood pressure.

3) Both high blood pressure and high body fat may be caused by a lurking variable, such as a genetic or lifestyle trait.

### 37. Reading.

a) The principal's description of a strong, positive trend is misleading. First of all, "trend" implies a change over time. These data were gathered during one year, at different grade levels. To observe a trend, one class's reading scores would have to be followed through several years. Second, the strong, positive relationship only indicates the yearly improvement that would be expected, as children get older. For example, the 4th graders are reading at approximately a 4th grade level, on average. This means that the school's students are progressing adequately in their reading, not extraordinarily. Finally, the use of average reading scores instead of individual scores increases the strength of the association.

b) The plot appears very straight. The correlation between grade and reading level is very high, probably between 0.9 and 1.0.

c) If the principal had made a scatterplot of all students' scores, the correlation would have likely been lower. Averaging reduced the scatter, since each grade level has only one point instead of many, which inflates the correlation.

**d)** If a student is reading at grade level, then that student's reading score should equal his or her grade level. The slope of that relationship is 1. That would be "acceptable", according to the measurement scale of reading level. Any slope greater than 1 would indicate above grade level reading scores, which would certainly be acceptable as well. A slope less than 1 would indicate below grade level average scores, which would be unacceptable.

## 39. Heating.

**a)** The model predicts a decrease in $2.13 in heating cost for an increase in temperature of 1° Fahrenheit. Generally, warmer months are associated with lower heating costs.

**b)** When the temperature is 0° Fahrenheit, the model predicts a monthly heating cost of $133.

**c)** When the temperature is around 32° Fahrenheit, the predictions are generally too high. The residuals are negative, indicating that the actual values are lower than the predicted values.

**d)**

$$\widehat{Cost} = 133 - 2.13(Temp)$$

$$\widehat{Cost} = 133 - 2.13(10)$$

$$\widehat{Cost} = \$111.70$$

According to the model, the heating cost in a month with average daily temperature 10° Fahrenheit is expected to be $111.70.

**e)** The residual for a 10° day is approximately –$6, meaning that the actual cost was $6 less than predicted, or $111.70 – $6 = $105.70.

**f)** The model is not appropriate. The residuals plot shows a definite curved pattern. The association between monthly heating cost and average daily temperature is not linear.

**g)** A change of scale from Fahrenheit to Celsius would not affect the relationship. Associations between quantitative variables are the same, no matter what the units.

## 41. TBill rates 2016.

**a)** $r = \sqrt{R^2} = \sqrt{0.776} = 0.881$. The correlation between rate and year is +0.881, since the scatterplot shows a positive association.

**b)** According to the model, treasury bill rates during this period increased at about 0.25% per year, starting from an interest rate of about 0.61% in 1950.

c) The linear regression equation predicting interest rate from year is

$$\widehat{Rate} = 0.61149 + 0.24788(Year - 1950)$$

$$\widehat{Rate} = 0.61149 + 0.24788(70)$$

$$\widehat{Rate} = 17.96$$

According to the model, the interest rate is predicted to be about 18% in the year 2020.

d) This prediction is not likely to have been a good one. Extrapolating 40 years beyond the final year in the data would be risky, and unlikely to be accurate.

**43. TBill rates 2014 revisited.**

a) Treasure bill rates peaked around 1980 and decreased afterward. This regression model has a negative slope and a high intercept.

b) The model that predicts the interest rate on 3-month Treasury bills from the number of years since 1950 is $\widehat{Rate} = 18.5922 - 0.29646(Year - 1950)$. This model predicts the interest rate to be –2.16%%, a negative rate, which doesn't make sense. This is much lower than the prediction made with the other model.

c) Even though we separated the data, there is no way of knowing if this trend will continue. And the rate cannot become negative, so we have clearly extrapolated far beyond what the data can support.

d) It is clear from the scatterplot that we can't count on TBill rates to change in a linear way over many years, since we have witnessed one change in direction already. It would not be wise to use any regression model to predict rates.

**45. Gestation.**

a) The association would be stronger if humans were removed. The point on the scatterplot representing human gestation and life expectancy is an outlier from the overall pattern and detracts from the association. Humans also represent an influential point. Removing the humans would cause the slope of the linear regression model to increase, following the pattern of the non-human animals much more closely.

b) The study could be restricted to non-human animals. This appears justifiable, since one could point to a number of environmental factors that could influence human life expectancy and gestation period, making them incomparable to those of animals.

c) The correlation is moderately strong. The model explains 72.2% of the variability in gestation period of non-human animals.

**d)** For every year increase in life expectancy, the model predicts an increase of approximately 15.5 days in gestation period.

**e)**

$$\widehat{Gest} = -39.5172 + 15.4980\, LifEx$$

$$\widehat{Gest} = -39.5172 + 15.4980\,(20)$$

$$\widehat{Gest} \approx 270.4428$$

According to the linear model, monkeys with a life expectancy of 20 years are expected to have gestation periods of about 270.5 days. Care should be taken when assessing the accuracy of this prediction. First of all, the residuals plot has not been examined, so the appropriateness of the model is questionable. Second, it is unknown whether or not monkeys were included in the original 17 non-human species studied. Since monkeys and humans are both primates, the monkeys may depart from the overall pattern as well.

**47. Elephants and hippos.**

**a)** Hippos are more of a departure from the pattern. Removing that point would make the association appear to be stronger.

**b)** The slope of the regression line would increase, pivoting away from the hippos point.

**c)** Anytime data points are removed, there must be a justifiable reason for doing so, and saying, "I removed the point because the correlation was higher without it" is not a justifiable reason.

**d)** Elephants are an influential point. With the elephants included, the slope of the linear model is 15.4980 days gestation per year of life expectancy. When they are removed, the slope is 11.6 days per year. The decrease is significant.

**49. Marriage age 2015 predictions.**

**a)** The linear model used to predict average female marriage age from year is: $\widehat{Age} = -112.543 + 0.068479\,Year$. The residuals plot shows a clear pattern. The model predicts that each year that passes is associated with an increase of 0.068479 years in the average female age at first marriage. The model predicts that the average female marriage age in 2025 will be approximately 26.13 years.

**b)** Don't place too much faith in this prediction. The residuals plot shows a clear pattern, indicating that this model is not appropriate. Additionally, this prediction is for a year that is 10 years higher than the highest year for which we have an average female marriage age. Extrapolation is risky when making predictions.

**c)** An extrapolation of more than 50 years into the future would be absurd. There is no reason to believe the trend would continue.

**d)** The linear model used to predict average female marriage age from year is: $\widehat{Age} = -274.742 + 0.149983\,Year$. The residuals plot shows a clear pattern. The model predicts that each year that passes is associated with an increase of 0.149983 years in the average female age at first marriage. The model predicts that the average female marriage age in 2025 will be approximately 28.97 years.

**e)** Don't place too much faith in this prediction. Though the residuals are quite small, the residuals plot shows a clear pattern, indicating that this model is not appropriate. Additionally, this prediction is for a year that is 10 years higher than the highest year for which we have an average female marriage age. Extrapolation is risky when making predictions. This model is better than the previous one, but still not a good model.

**f)** An extrapolation of more than 50 years into the future would be absurd. There is no reason to believe the trend would continue. Again, this would be an unreasonable extrapolation.

**51. Fertility and life expectancy 2014.**

**a)** The association between fertility rate and life expectancy is moderate, linear, and negative. The residual plot is reasonably scattered with no evidence of nonlinearity, so we can fit the regression model. But there seems to be an outlier, which could be affecting the regression model.

**b)** There is one outlier, Niger, with a higher life expectancy than typical for its large family size.

**c)** $R^2 = 63.9\%$, so 63.9% of the variability in life expectancy is explained by variability in the fertility rate.

**d)** The government leaders should not suggest that women have fewer children in order to raise the life expectancy. Although there is evidence of an association between the fertility rate and life expectancy, this does not mean that one causes the other. There may be lurking variables involved, such as economic conditions, social factors, or level of health care.

## 53. Inflation 2016.

a) The trend in Consumer Price Index is strong, non-linear, and positive. Generally, CPI has increased over the years, but the rate of increase has become much greater since approximately 1971. Other characteristics include fluctuations in CPI in the years prior to 1950.

b) Answers may vary. In order to effectively predict the CPI over the next decade, use only the most recent trend. The trend since 1971 is straight enough to apply the linear model. Prior to 1971, the trend is radically different from that of recent years, and is of no use in predicting CPI for the next decade.

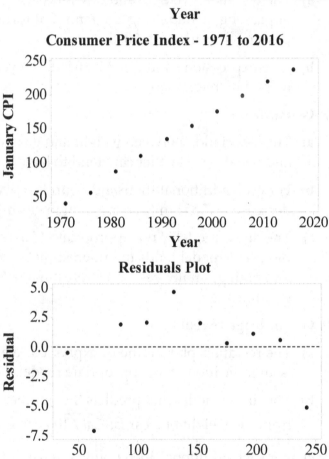

The linear model that predicts CPI from year is

$\widehat{CPI} = -8797.47 + 4.48388\,Year$

$R^2 = 99.7\%$, meaning that the model predicts 99.7% of the variability in CPI.

The residuals plot shows some pattern, but the residuals are small, so the linear model is appropriate.

According to the model, the CPI is expected to increase by $4.48 each year, for 1971 − 2016.

$$\widehat{CPI} = -8797.47 + 4.48388\,Year$$

$$\widehat{CPI} = -8797.47 + 4.48388(2026)$$

$$\widehat{CPI} = 286.87$$

As with any model, care should be taken when extrapolating. If the pattern continues, the model predicts that the CPI in 2026 will be approximately $286.87.

### 55. Oakland passengers 2016 revisited.

a) The residuals cycle up and down because there is a yearly pattern in the number of passengers departing Oakland, California. There is also a sudden decrease in passenger traffic after 2008.

b) A re-expression should not be tried. A cyclic pattern such as this one cannot be helped by re-expression.

### 57. Gas mileage.

a) The association between weight and gas mileage of cars is fairly linear, strong, and negative. Heavier cars tend to have lower gas mileage.

b) For each additional thousand pounds of weight, the linear model predicts a decrease of 7.652 miles per gallon in gas mileage.

c) The linear model is not appropriate. There is a curved pattern in the residuals plot. The model tends to underestimate gas mileage for cars with relatively low and high gas mileages, and overestimates the gas mileage of cars with average gas mileage.

### 59. Gas mileage revisited.

a) The residuals plot for the re-expressed relationship is much more scattered. This is an indication of an appropriate model.

b) The linear model that predicts the number of gallons per 100 miles in gas mileage from the weight of a car is: $\widehat{Gal/100} = 0.625 + 1.178(Weight)$.

c) For each additional 1000 pounds of weight, the model predicts that the car will require an additional 1.178 gallons to drive 100 miles.

d)

$$\widehat{Gal/100} = 0.625 + 1.178(Weight)$$

$$\widehat{Gal/100} = 0.625 + 1.178(3.5)$$

$$\widehat{Gal/100} = 4.748$$

According to the model, a car that weighs 3500 pounds (3.5 thousand pounds) is expected to require approximately 4.748 gallons to drive 100 miles, or 0.04748 gallons per mile.

This is $\dfrac{1}{0.04748} \approx 21.06$ miles per gallon.

### 61. USGDP 2016.

**a)** Although nearly 97% of the variation in GDP can be accounted for by the model, the residuals plot should be examined to determine whether or not the model is appropriate.

**b)** This is not a good model for these data. The residuals plot shows curvature.

### 63. Better GDP model?

There is still a pattern in the residuals. This much pattern still indicates an inappropriate model. Since re-expressing with logarithms went too far, changing the curvature in the other direction, we should move up the ladder of powers. A square root re-expression might straighten the plot.

### 65. Brakes.

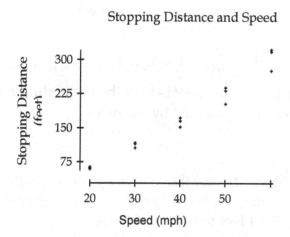

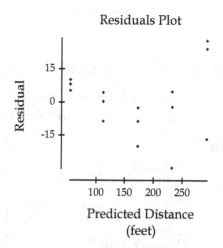

**a)** The association between speed and stopping distance is strong, positive, and appears straight. Higher speeds are generally associated with greater stopping distances. The linear regression model, with equation

$$\widehat{Distance} = -65.9 + 5.98\left(Speed\right),$$ has $R^2 = 96.9\%$, meaning that the model explains 96.9% of the variability in stopping distance. However, the residuals plot has a curved pattern. The linear model is not appropriate. A model using re-expressed variables should be used.

**b)** Stopping distances appear to be relatively higher for higher speeds. This increase in the rate of change might be able to be straightened by taking the square root of the response variable, stopping distance. The scatterplot of Speed versus $\sqrt{Distance}$ seems like it might be a bit straighter.

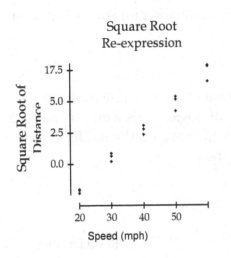

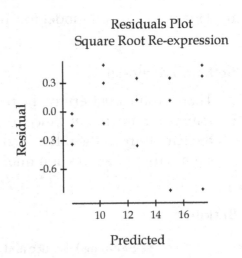

**c)** The model for the re-expressed data is $\overline{\sqrt{Distance}} = 3.303 + 0.235(Speed)$. The residuals plot shows no pattern, and $R^2 = 98.4\%$, so 98.4% of the variability in the square root of the stopping distance can be explained by the model.

**d)**

$$\overline{\sqrt{Distance}} = 3.303 + 0.235(Speed)$$

$$\overline{\sqrt{Distance}} = 3.303 + 0.235(55)$$

$$\overline{\sqrt{Distance}} = 16.228$$

$$\overline{Distance} = 16.228^2 \approx 263.4$$

According to the model, a car traveling 55 mph is expected to require approximately 263.4 feet to come to a stop.

**e)**

$$\overline{\sqrt{Distance}} = 3.303 + 0.235(Speed)$$

$$\overline{\sqrt{Distance}} = 3.303 + 0.235(70)$$

$$\overline{\sqrt{Distance}} = 19.753$$

$$\overline{Distance} = 19.753^2 \approx 390.2$$

According to the model, a car traveling 70 mph is expected to require approximately 390.2 feet to come to a stop.

**f)** The level of confidence in the predictions should be quite high. $R^2$ is high, and the residuals plot is scattered. The prediction for 70 mph is a bit of an extrapolation, but should still be reasonably close.

**67. Baseball salaries 2016.**

a) The association between year and highest salary is curved, strong, and positive. Salaries were flat for many years, and began to increase in the late 1970s, then increased more rapidly as in recent years. The trend is not linear and the plot is not straight enough for a regression model.

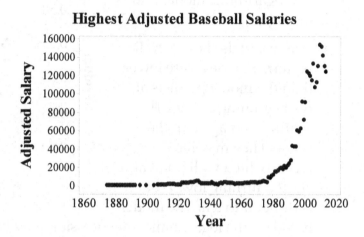

**Highest Adjusted Baseball Salaries**

b) Re-expression using the logarithm of the adjusted salaries straightens the plot significantly.

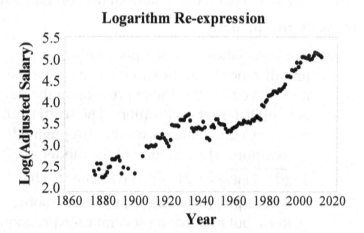

**Logarithm Re-expression**

c)
```
Model Summary

        S     R-sq   R-sq(adj)   R-sq(pred)
 0.278834   87.68%      87.58%       87.32%

Coefficients

Term           Coef   SE Coef   T-Value   P-Value   VIF
Constant   -32.3897      1.19    -27.15     0.000
Year       0.018492  0.000613     30.18     0.000   1.00
```

The equation of the model is $\log(\widehat{AdjSalary}) = -32.3897 + 0.018492(Year)$. Salaries have been increasing on a log scale by about .018 logmillion \$/year. Transforming back to dollars, that's a bit more than a million dollars/year.

**d)** The plot of the residuals for the logarithmic model is at the right.

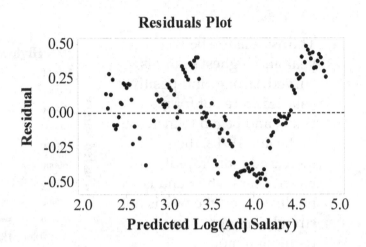

**Residuals Plot**

**e)** The residuals show a cyclic pattern. Salaries were lower than the model predicts at the beginning of the 20th century and again in the 1960s. They may have recently hit a high point and started turning lower, but it is difficult to tell. The model based on the logarithmic re-expression may be the best model we can find, but it doesn't explain the pattern in highest baseball salary over time.

**69. Is Pluto a planet?**

**a)** The association between planetary position and distance from the sun is strong, positive, and curved. A good re-expression of the data is Log(Distance) vs. Position. The scatterplot with regression line shows the straightened association. The equation of the model is $\overline{\log(Distance)} = 1.246 + 0.271(Position)$. The residuals plot (below right) may have some pattern, but after trying several re-expressions, this is the best that can be done. $R^2 = 98.1\%$, so the model explains 98.1% of the variability in the log of the planet's distance from the sun.

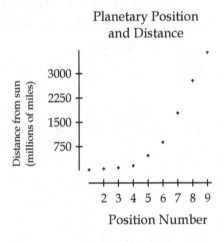

Planetary Position and Distance

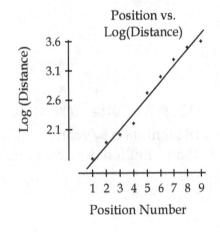

Position vs. Log(Distance)

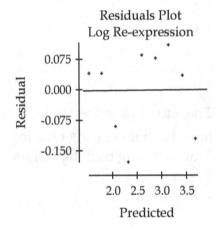

Residuals Plot Log Re-expression

**b)** At first glance, this model appears to provide little evidence to support the contention of the International Astronomical Union. Pluto appears to fit the pattern, although Pluto's distance from the sun is a bit less than expected. A model generated without Pluto does not have a dramatically improved residuals plot, does not have a significantly higher $R^2$, nor a different slope. Pluto does not appear to be influential.

But don't forget that a logarithmic scale is being used for the vertical axis. The higher up the vertical axis you go, the greater the effect of a small change.

$$\overline{\log(Distance)} = 1.245807 + 0.270792(Position)$$

$$\overline{\log(Distance)} = 1.245807 + 0.270792(9)$$

$$\overline{\log(Distance)} = 3.682935$$

$$\overline{Distance} = 10^{3.682935} \approx 4819$$

According to the model, the 9th planet in the solar system is predicted to be approximately 4819 million miles away from the sun. Pluto is actually 3672 million miles away.

Pluto doesn't fit the pattern for position and distance in the solar system. In fact, the model made with Pluto included isn't a good one, because Pluto influences those predictions.

The model without Pluto, $\overline{\log(Distance)} = 1.203650 + 0.283439(Position)$, works much better. It has a high $R^2$, and scattered residuals plot. This new model predicts that the 9th planet should be a whopping 5683 million miles away from the sun! There is evidence that the IAU is correct. Pluto doesn't behave like a planet in its relation to position and distance.

**71. Planets, and Eris.**

A planet tenth from the sun (with the asteroid belt as a failed planet and Pluto not included as a planet) is predicted to be about 4714 million miles away from the sun.

$$\overline{\log(Distance)} = 1.28514 + 0.238826(Position)$$

$$\overline{\log(Distance)} = 1.28514 + 0.238826(10)$$

$$\overline{\log(Distance)} = 3.6734$$

$$\overline{Distance} = 10^{3.6734} \approx 4714$$

This distance is much shorter than the actual distance of Eris, about 6300 miles.

Similarly, with the asteroid belt as a failed planet, and Pluto as the tenth planet, Eris, as the eleventh planet, is predicted to be 7368 million miles away, much farther away than it actually is. Like Pluto,

$$\overline{\log(Distance)} = 1.306310 + 0.232822(Position)$$

$$\overline{\log(Distance)} = 1.306310 + 0.232822(11)$$

$$\overline{\log(Distance)} = 3.867352$$

$$\overline{Distance} = 10^{3.867352} \approx 7368$$

Eris doesn't behave like a planet in its relation to position and distance.

**73. Logs (not logarithms).**

a) The association between the diameter of a log and the number of board feet of lumber is strong, positive, and curved. As the diameter of the log increases, so does the number of board feet of lumber contained in the log.

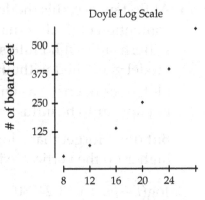

Doyle Log Scale

The model used to generate the table used by the log buyers is based upon a square root re-expression. The values in the table correspond exactly to the model

$$\overline{\sqrt{BoardFeet}} = -4 + Diameter .$$

b)

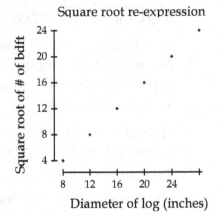

Square root re-expression

$$\overline{\sqrt{BoardFeet}} = -4 + Diameter$$

$$\overline{\sqrt{BoardFeet}} = -4 + (10)$$

$$\overline{\sqrt{BoardFeet}} = 6$$

$$\overline{BoardFeet} = 36$$

According to the model, a log 10" in diameter is expected to contain 36 board feet of lumber.

c)

$$\overline{\sqrt{BoardFeet}} = -4 + Diameter$$

$$\overline{\sqrt{BoardFeet}} = -4 + (36)$$

$$\overline{\sqrt{BoardFeet}} = 32$$

$$\overline{BoardFeet} = 1024$$

According to the model, a log 36" in diameter is expected to contain 1024 board feet of lumber.

Normally, we would be cautious of this prediction, because it is an extrapolation beyond the given data, but since this is a prediction made from an exact model based on the volume of the log, the prediction will be accurate.

### 75. Life expectancy history.

The association between year and life expectancy is strong, curved and positive. As the years passed, life expectancy for white males has increased.

The linear model,
$\widehat{LifeExp} = 46.57 + 2.68(Dec)$,
explains 95.0% of the variability in the life expectancy of white males, but has a residuals plot that reveals a strong pattern.

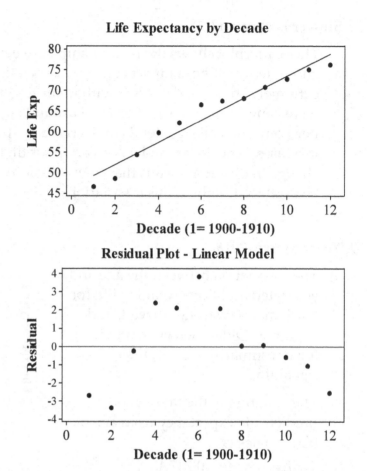

The association between Log(*Year*) and Log(*Life Expectancy*) is strong, positive, and reasonably straight. The model is $\widehat{\log_{10}(LifeExp)} = 1.646 + 0.215(\log_{10}(Year))$.

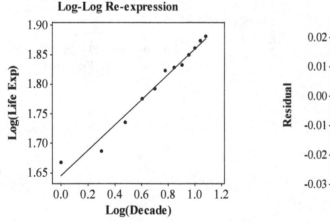

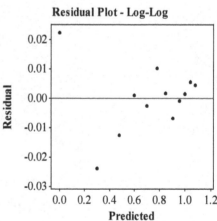

This model explains 97.4% of the variability in the logarithm of the Life Expectancy. The residuals plot shows some pattern, but seems more scattered than the residuals plot for the linear model.

### 77. Slower is cheaper?

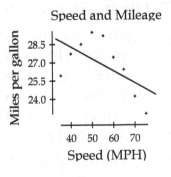

Speed and Mileage

The scatterplot shows the relationship between speed and mileage of the compact car. The association is extremely strong and curved, with mileage generally increasing as speed increases, until around 50 miles per hour, then mileage tends to decrease as speed increases. The linear model is a very poor fit, but the change in direction means that re-expression cannot be used to straighten this association.

### 79. Years to live, 2016.

a)  The association between the age and estimated additional years of life for black males is strong, curved, and negative. Older men generally have fewer estimated years of life remaining.

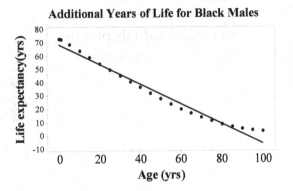

The equation of the linear model that predicts life expectancy from age of black males is

$$\widehat{LifeExp} = 68.63 - 0.747\,Age.$$

The model is not a good fit. The residuals plot shows a curved pattern.

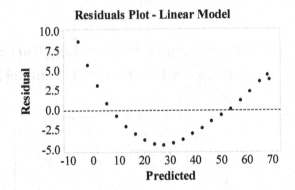

**b)** Answers may vary. The square root re-expression straightens the data considerably, but has an extremely patterned residuals plot. The model is not a mathematically appropriate model, but fits so closely that it should be fine for predictions within the range of data. The model explains 99.8% of the variability in the estimated number of additional years.

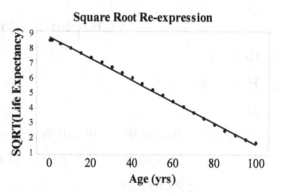

**Square Root Re-expression**

The equation of the square root model is $\sqrt{LifeExp} = 8.722 - 0.0721\,Age$ .

$$\widehat{\sqrt{LifeExp}} = 8.722 - 0.0721\,Age$$

$$\widehat{\sqrt{LifeExp}} = 8.722 - 0.0721(18)$$

$$\widehat{\sqrt{LifeExp}} = 7.4242$$

$$\widehat{LifeExp} = 7.4242^2 \approx 55.12$$

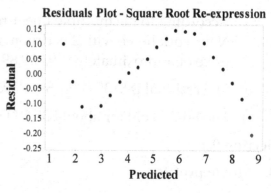

**Residuals Plot - Square Root Re-expression**

According to the model, an 18-year-old black male is expected to have 55.12 years of life remaining, and live to be approximately 73.12 years of age.

**c)** The residuals plot is extremely patterned, so the model is not appropriate. However, the residuals are very small, making for a tight fit. Since 18 years is within the range of the data, the prediction should be at least reasonable.

# Chapter 9 – Multiple Regression

## Section 9.1

1. **House prices.**

   **a)**

   $$\widehat{Price} = 20{,}986.09 - 7483.10\, Bedrooms + 93.84\, LivingArea$$
   $$= 20{,}986.09 - 7483.10(2) + 93.84(1000)$$
   $$= 99{,}859.89$$

   According to the multiple linear regression model, we would expect an Upstate New York home with 2 bedrooms with 1000 square feet of living space to have a price of approximately $99,859.89.

   **b)** The residual is $135,000 – $99,859.89 = $35,140.11.

   **c)** The house sold for about $35,100 more than our estimate.

## Section 9.2

3. **Movie profits.**

   **a)** $\widehat{USGross} = -52.3692 + 0.9723\, Budget + 0.3872\, RunTime + 0.6403\, CriticsScore$

   **b)** After allowing for the effects of *RunTime* and *CriticsScore*, each million dollars spent making a film yields about 0.9723 million dollars in gross revenue.

## Section 9.3

5. **More movies profits.**

   **a)** Linearity: The plot is reasonably linear with no bends.

   **b)** Equal spread: The plot fails the Equal Spread condition. It is much more spread out to the right than on the left.

   **c)** Normality: A scatterplot of two of the variables doesn't tell us anything about the distribution of the residuals.

## Section 9.4

7. **Movie profits once more.**

   **a)** The partial regression plot is for US Gross residual and Budget residual, so the slope of the line is the Budget coefficient from the multiple regression, 0.9723.

   **b)** *Avatar* has a budget and US gross that is influential in the regression, and pulls the line toward it. If *Avatar*'s point were removed, the slope of the line would decrease.

**Section 9.5**

9. **Indicators.**

   **a)** Use an indicator. Code Male = 0, Female = 1 (or the reverse).

   **b)** Treat it as a quantitative predictor.

   **c)** Use an indicator. Code older than 65 = 1, Younger than 65 = 0.

11. **Interpretations.**

   **a)** There are two problems with this interpretation. First, the other predictors are not mentioned. Secondly, the prediction should be stated in terms of a mean, not a precise value.

   **b)** This is a correct interpretation.

   **c)** This interpretation attempts to predict in the wrong direction. This model cannot predict *lotsize* from *price*.

   **d)** $R^2$ concerns the fraction of variability accounted for by the regression model, not the fraction of data values.

13. **Predicting final exams.**

   **a)** $\widehat{Final} = -6.7210 + 0.2560(Test1) + 0.3912(Test2) + 0.9015(Test3)$

   **b)** $R^2 = 77.7\%$, which means that 77.7% of the variation in *Final* grade is accounted for by the multiple regression model.

   **c)** According to the multiple regression model, each additional point on *Test3* is associated with an average increase of 0.9015 points on the final, for students with given *Test1* and *Test2* scores.

   **d)** Test scores are probably collinear. If we are only concerned about predicting the final exam score, *Test1* may not add much to the regression. However, we would expect it to be associated with the final exam score.

15. **Attendance 2016.**

   **a)** *Won* and *Runs* are probably correlated. Including *Won* in the model is then very likely to change the coefficient of *Runs*, which now must be interpreted after allowing for the effects of *Won*.

   **b)** The Indians' actual *Attendance* was less than we would predict for the number of *Runs* they scored.

## 17. Home prices.

a) $\widehat{Price} = -152,037 + 9530 Bathrooms + 139.87 LivingArea$

b) According to the multiple regression model, the asking *Price* increases, on average, by about $139.87 for each additional square foot, for homes with the same number of bathrooms.

c) The number of bathrooms is probably correlated with the size of the house, even after considering the square footage of the bathroom itself. This correlation may account for the coefficient of *Bathrooms* not being discernibly different from 0. Moreover, the regression model does not predict what will happen when a house is modified, for example, by converting existing space into a bathroom.

## 19. Predicting finals II.

**Straight enough condition:** The plot of residuals versus fitted values looks curved, rising in the middle, and falling on both ends. This is a potential difficulty.
**Randomization condition:** It is reasonable to think of this class as a representative sample of all classes.
**Nearly Normal condition:** The Normal probability plot and the histogram of residuals suggest that the highest five residuals are extraordinarily high.
**Does the plot thicken? condition:** The spread is not consistent over the range of predicted values.

These data may benefit from a re-expression.

## 21. Admin performance.

a)
$$\widehat{Salary} = 9.788 + 0.11 Service + 0.053 Education$$
$$+ 0.071 Score + 0.004 Speed + 0.065 Dictation$$

b) $\widehat{Salary} = 9.788 + 0.11(120) + 0.053(9) + 0.071(50) + 0.004(60) + 0.065(30) \approx \$29,200$

c) Although *Age* and *Salary* are positively correlated, after removing the effects of years of education and months of service from *Age*, what is left is years not spent in education or service. Those non-productive years may well have a negative effect on salary.

## 23. Body fat revisited.

a) According to the linear model, each pound of *Weight* is associated with a 0.189 increase in *%Body Fat*.

**b)** After removing the linear effects of *Waist* and *Height*, each pound of *Weight* is associated, on average, with a decrease of 0.10% in *%Body Fat*. The change in coefficient and sign is a result of including the other predictors. We expect *Weight* to be correlated with both *Waist* and *Height*. It may be collinear with them.

**c)** We should examine the partial regression plot for *Weight*. It would have an *x*-axis that is *Weight* with the effects of *Waist* and *Height* removed, so we can understand that better.

**25. Breakfast cereals again.**

**a)** $\widehat{Calories} = 83.0469 + 0.05721\,Sodium - 0.01933\,Potassium + 2.38757\,Sugars$

**b)** We should examine a scatterplot of residuals vs. predicted values and a Normal probability plot of the residuals.

**c)** No, adding *Potassium* wouldn't necessarily lower the *Calories* in a breakfast cereal. A regression model doesn't predict what would happen if we change a value of a predictor. It only models the data as they are.

**27. Hand dexterity.**

**a)** For children of the same *Age*, their *Dominant Hand* was faster on this task by about 0.304 seconds on average.

**b)** Yes. The relationship between *Speed* and *Age* is straight and the lines for dominant and non-dominant hands are very close to parallel.

**29. Scottish hill races, men and women.**

**a)** The change in *Time* with *Distance* is different for men and women, so a different slope is needed in the model.

**b)** After accounting for *Distance*, the increase in race time with distance is greater for men than for women. Said another way, women's times increase less rapidly for longer races than do men's.

## Review of Part II – Exploring Relationships Between Variabless

**R2.1. College.**

| | |
|---|---|
| % over 50: $r = 0.69$ | The only moderate, positive correlation in the list. |
| % under 20: $r = -0.71$ | Moderate, negative correlation ($-0.98$ is too strong) |
| % Full-time Fac.: $r = 0.09$ | No correlation. |
| % Gr. on time: $r = -0.51$ | Moderate, negative correlation (not as strong as |
| %under 20) | |

**R2.3. Vineyards, more information.**

a) There does not appear to be an association between ages of vineyards and the price of products. $r = \sqrt{R^2} = \sqrt{0.027} = 0.164$, indicating a very weak association, at best. The model only explains 2.7% of the variability in case price. Furthermore, the regression equation appears to be influenced by two outliers, products from vineyards over 30 years old, with relatively high case prices.

b) This analysis tells us nothing about vineyards worldwide. There is no reason to believe that the results for the Finger Lakes region are representative of the vineyards of the world.

c) The linear equation used to predict case price from age of the vineyard is:
$$\widehat{CasePrice} = 92.765 + 0.567284\,Years$$

d) This model is not useful because only 2.7% of the variability in case price is accounted for by the ages of the vineyards. Furthermore, the slope of the regression line seems influenced by the presence of two outliers, products from vineyards over 30 years old, with relatively high case prices.

**R2.5.** **Twins by year 2014.**

a) The association between year and the twin birth rate is strong, positive, and appears non-linear. Generally, the rate of twin births has increased over the years. The linear model that predicts the rate of twin births from the year is:

$$\widehat{Twins} = 17.77 + 0.551(Years\,Since\,1980)$$

**Twin Birth Rate - 1980 to 2009**

b) For each year that passes, the model predicts that the twin birth rate will increase by an average of approximately 0.55 twin births per 1000 live births.

c) $\widehat{Twins} = 17.77 + 0.551(Years\,Since\,1980)$

$\widehat{Twins} = 17.77 + 0.551(34)$

$\widehat{Twins} = 36.504$

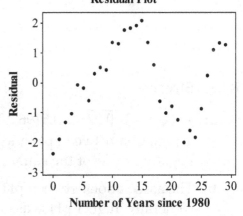

**Residual Plot**

According to the model, the twin birth rate is expected to be 36.50 twin births per 1000 live births in the US in 2014. This is reasonably close to the actual rate of 33.9 twin births per 1000 women, even though this was an extrapolation based on a model with a highly patterned residuals plot.
However, being close in hindsight is not justification for using this model to make this prediction.

**d)** The linear model that predicts the rate of twin births from the year is:

$$\widehat{Twins} = 18.21 + 0.508\,(Years\ Since\ 1980)$$

This model fits very well, with $R^2 = 97\%$. However, the residuals plot still shows the fluctuating pattern that troubled us before. There may be more to understand about twin births than is available from these data.

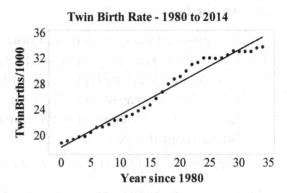

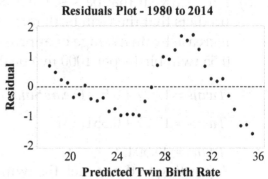

**R2.7. Streams.**

**a)** $r = \sqrt{R^2} = \sqrt{0.27} = -0.5196$. The association between pH and BCI appears negative in the scatterplot, so use the negative value of the square root.

**b)** The association between pH and BCI is negative, moderate, and linear. Generally, higher pH is associated with lower BCI. Additionally, BCI appears more variable for higher values of pH.

**c)** In a stream with average pH, the BCI would be expected to be average, as well.

**d)** In a stream where the pH is 3 standard deviations above average, the BCI is expected to be 1.56 standard deviations below the mean level of BCI. ($r\,(3) = -0.5196(3) = -1.56$)

**R2.9. Streams II.**

**a)** The association between pH and BCI is negative, moderate, and linear. Generally, higher pH is associated with lower BCI. The scatterplot (shown in the previous exercise) is straight enough to find a linear regression model.

The linear regression model is: $\widehat{BCI} = 2733.37 - 197.694\,pH$.

**b)** The model predicts that BCI decreases by an average of 197.69 point per point of pH.

**c)** The model predicts that a stream with pH of 8.2 will have a BCI of approximately 1112.3.

$$\widehat{BCI} = 2733.37 - 197.694\,pH$$
$$\widehat{BCI} = 2733.37 - 197.694\,(8.2)$$
$$\widehat{BCI} = 1112.2792$$

**R2.11. Streams III.**

a) The regression model that predicts *BCI* from *pH* and water hardness is
$\widehat{BCI} = 2342.95 - 137.833pH - 0.337210Hard$ .

b) This model is slightly better than the model in the prior exercise. $R^2$ is slightly higher and *s* is slightly lower.

c) According to the multiple linear regression model, a stream with a *pH* of 8.2 and a hardness value of 205 is predicted to have *BCI* of approximately 1144.

$\widehat{BCI} = 2342.95 - 137.833pH - 0.337210Hard$

$\widehat{BCI} = 2342.95 - 137.833(8.2) - 0.337210(205)$

$\widehat{BCI} = 1143.59135$

d) $1309 - 1144 = 165$. This is the residual.

e) After allowing for the effects of *Hardness* of the water, *BCI* decreases by about 137.8 per unit of *pH*.

**R2.13. Traffic.**

a)

$b_1 = r\dfrac{s_y}{s_x}$

$-0.352 = r\dfrac{9.68}{27.07}$

$r = -0.984$

The correlation between traffic density and speed is $r = -0.984$

b) $R^2 = (-0.984)^2 = 0.969$.
The variation in the traffic density accounts for 96.9% of the variation in speed.

c)

$\widehat{Speed} = 50.55 - 0.352\,Density$

$\widehat{Speed} = 50.55 - 0.352(50)$

$\widehat{Speed} = 32.95$

According to the linear model, when traffic density is 50 cars per mile, the average speed of traffic on a moderately large city thoroughfare is expected to be 32.95 miles per hour.

d)

$\widehat{Speed} = 50.55 - 0.352\,Density$

$\widehat{Speed} = 50.55 - 0.352(56)$

$\widehat{Speed} = 30.84$

According to the linear model, when traffic density is 56 cars per mile, the average speed of traffic on a moderately large city thoroughfare is expected to be 30.84 miles per hour. If traffic is actually moving at 32.5 mph, the residual is $32.5 - 30.84 = 1.66$ miles per hour.

**e)**

$$\widehat{Speed} = 50.55 - 0.352\,Density$$

$$\widehat{Speed} = 50.55 - 0.352\,(125)$$

$$\widehat{Speed} = 6.55$$

According to the linear model, when traffic density is 125 cars per mile, the average speed of traffic on a moderately large city thoroughfare is expected to be 6.55 miles per hour. The point with traffic density 125 cars per minute and average speed 55 miles per hour is considerably higher than the model would predict. If this point were included in the analysis, the slope would increase.

**f)** The correlation between traffic density and average speed would become weaker. The influential point (125, 55) is a departure from the pattern established by the other data points.

**g)** The correlation would not change if kilometers were used instead of miles in the calculations. Correlation is a "unitless" measure of the degree of linear association based on $z$-scores, and is not affected by changes in scale. The correlation would remain the same, $r = -0.984$.

**R2.15. Cars, correlations.**

**a)** Weight, with a correlation of –0.903, seems to be most strongly associated with fuel economy, since the correlation has the largest magnitude (distance from zero). However, without looking at a scatterplot, we can't be sure that the relationship is linear. Correlation might not be an appropriate measure of the strength of the association if the association is non-linear.

**b)** The negative correlation between weight and fuel economy indicates that, generally, cars with higher weights tend to have lower mileages than cars with lower weights. Once again, this is only correct if the association between weight and fuel economy is linear.

**c)** $R^2 = (-0.903)^2 = 0.815$. The variation in weight accounts for 81.5% of the variation in mileage. Once again, this is only correct if the association between weight and fuel economy is linear.

**R2.17. Cars, horsepower.**

**a)** The linear model that predicts the horsepower of an engine from the weight of the car is: $\widehat{Horsepower} = 3.49834 + 34.3144\,Weight$.

**b)** The weight is measured in thousands of pounds. The slope of the model predicts an increase of about 34.3 horsepower for each additional unit of weight. 34.3 horsepower for each additional thousand pounds makes more sense than 34.3 horsepower for each additional pound.

c) Since the residuals plot shows no pattern, the linear model is appropriate for predicting horsepower from weight.

d)

$$\widehat{Horsepower} = 3.49843 + 34.3144\,Weight$$

$$\widehat{Horsepower} = 3.49843 + 34.3144\,(2.595)$$

$$\widehat{Horsepower} \approx 92.544$$

According to the model, a car weighing 2595 pounds is expected to have 92.543 horsepower. The actual horsepower of the car is: $92.544 + 22.5 \approx 115.0$ horsepower.

### R2.19. Cars, more efficient?

a) The residual plot is curved, suggesting that the conditions for inference are not satisfied. Specifically, the Straight Enough condition is not satisfied.

b) The reciprocal model is preferred to the model in the previous exercise. The new residual plot has no pattern, and appears to satisfy the conditions for inference. There is, however, one outlier to consider.

### R2.21. Old Faithful again.

a) The association between the duration of eruption and the interval between eruptions of Old Faithful is fairly strong, linear, and positive. Long eruptions are generally associated with long intervals between eruptions. There are also two distinct clusters of data, one with many short eruptions followed by short intervals, the other with many long eruptions followed by long intervals, with only a few medium eruptions and intervals in between.

b) The linear model used to predict the interval between eruptions is:

$$\widehat{Interval} = 33.9668 + 10.3582\,Duration.$$

c) As the duration of the previous eruption increases by one minute, the model predicts an increase of about 10.4 minutes in the interval between eruptions.

d) $R^2 = 77.0\%$, so the model accounts for 77% of the variability in the interval between eruptions. The predictions should be fairly accurate, but not precise. Also, the association appears linear, but we should look at the residuals plot to be sure that the model is appropriate before placing too much faith in any prediction.

e)

$$\widehat{Interval} = 33.9668 + 10.3582\,Duration$$

$$\widehat{Interval} = 33.9668 + 10.3582\,(4)$$

$$\widehat{Interval} \approx 75.4$$

According to the model, if an eruption lasts 4 minutes, the next eruption is expected to occur in approximately 75.4 minutes.

f) The actual eruption at 79 minutes is 3.6 minutes later than predicted by the model. The residual is $79 - 75.4 = 3.6$ minutes. In other words, the model under-predicted the interval.

## R2.23. How old is that tree?

**a)** The correlation between tree diameter and tree age is $r = 0.888$. Although the correlation is moderately high, this does not suggest that the linear model is appropriate. We must look at a scatterplot in order to verify that the relationship is straight enough to try the linear model. After finding the linear model, the residuals plot must be checked. If the residuals plot shows no pattern, the linear model can be deemed appropriate.

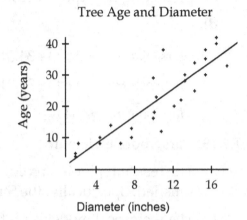

Tree Age and Diameter

**b)** The association between diameter and age of these trees is fairly strong, somewhat linear, and positive. Trees with larger diameters are generally older.

**c)** The linear model that predicts age from diameter of trees is:

$\widehat{Age} = -0.974424 + 2.20552\,Diameter$. This model explains 78.9% of the variability in age of the trees.

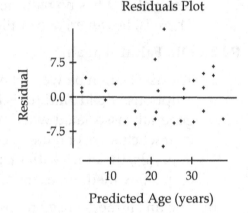

Residuals Plot

**d)** The residuals plot shows a curved pattern, so the linear model is not appropriate. Additionally, there are several trees with large residuals.

**e)** The largest trees are generally above the regression line, indicating a positive residual. The model is likely to underestimate these values.

## R2.25. Big screen.

TV screen sizes might vary from 19 to 70 inches. A TV with a screen that was 10 inches larger would be predicted to cost $10(0.03) = +0.3$, $10(0.3) = +3$, $10(3) = +30$, or $10(30) = +300$. Notice that the TV costs are measure in hundreds of dollars, so the potential price changes for getting a TV 10 inches larger are $30, $300, $3000, and $30,000. Only $300 is reasonable, so the slope must be 0.3.

## R2.27. No smoking?

**a)** The model from the previous exercise is for predicting the percent of expectant mothers who smoked during their pregnancies from the year, not the year from the percent.

**b)** The model that predicts the year from the percent of expectant mothers who smoked during pregnancy is: $\widehat{Year} = 2027.13 - 1.98233(\%)$. This model predicts that 0% of mothers will smoke during pregnancy in $2027.13 - 1.98233(0) \approx 2027$.

**c)** The lowest data point corresponds to 9.0% of expectant mothers smoking during pregnancy in 2011. The prediction for 0% is an extrapolation outside the scope of the data. There is no reason to believe that the model will be accurate at that point.

**R2.29. U.S. Cities.**

There is a strong, roughly linear, negative association between mean January temperature and latitude. U.S. cities with higher latitudes generally have lower mean January temperatures. There are two outliers, cities with higher mean January temperatures than the pattern would suggest.

**R2.31. Winter in the city.**

**a)** $R^2 = (-0.848)^2 \approx 0.719$. The variation in latitude explains 71.9% of the variability in average January temperature.

**b)** The negative correlation indicates that the as latitude increases, the average January temperature generally decreases.

**c)**
$$b_1 = r\frac{s_y}{s_x}$$
$$b_1 = (-0.848)\frac{13.49}{5.42}$$
$$b_1 = -2.1106125$$

$$\hat{y} = b_0 + b_1 x$$
$$\bar{y} = b_0 + b_1\bar{x}$$
$$26.44 = b_0 - 2.1106125(39.02)$$
$$b_0 = 108.79610$$

The equation of the linear model for predicting January temperature from latitude is: $\widehat{JanTemp} = 108.796 - 2.111\,Latitude$

**d)** For each additional degree of latitude, the model predicts a decrease of approximately 2.1°F in average January temperature.

**e)** The model predicts that the mean January temperature will be approximately 108.8°F when the latitude is 0°. This is an extrapolation, and may not be meaningful.

**f)**

$$\widehat{JanTemp} = 108.796 - 2.111\,Latitude$$
$$\widehat{JanTemp} = 108.796 - 2.111(40)$$
$$\widehat{JanTemp} \approx 24.4$$

According to the model, the mean January temperature in Denver is expected to be 24.4°F.

**g)** In this context, a positive residual means that the actual average temperature in the city was higher than the temperature predicted by the model. In other words, the model underestimated the average January temperature.

**R2.33. Olympic Jumps 2016.**

**a)** The association between Olympic long jump distances and high jump heights is strong, linear, and positive. Years with longer long jumps tended to have higher high jumps. There is one departure from the pattern. The year in which the Olympic gold medal long jump was the longest had a shorter gold medal high jump than we might have predicted.

**b)** There is an association between long jump and high jump performance, but it is likely that training and technique have improved over time and affected both jump performances.

**c)** The correlation would be the same, 0.910. Correlation is a measure of the degree of linear association between two quantitative variables and is unaffected by changes in units.

**d)** In a year when the high jumper jumped one standard deviation better than the average jump, the long jumper would be predicted to jump $r = 0.910$ standard deviations above the average long jump.

**R2.35. French.**

**a)** Most of the students would have similar weights. Regardless of their individual French vocabularies, the correlation would be near 0.

**b)** There are two possibilities. If the school offers French at all grade levels, then the correlation would be positive and strong. Older students, who typically weigh more, would have higher scores on the test, since they would have learned more French vocabulary. If French is not offered, the correlation between weight and test score would be near 0. Regardless of weight, most students would have horrible scores.

**c)** The correlation would be near 0. Most of the students would have similar weights and vocabulary test scores. Weight would not be a predictor of score.

**d)** The correlation would be positive and strong. Older students, who typically weigh more, would have higher test scores, since they would have learned more French vocabulary.

## R2.37. Lunchtime.

The association between time spent at the table and number of calories consumed by toddlers is moderate, roughly linear, and negative. Generally, toddlers who spent a longer time at the table consumed fewer calories than toddlers who left the table quickly. The scatterplot between time at the table and calories consumed is straight enough to justify the use of the linear model. The linear model that predicts the time number of calories consumed by a toddler from the time spent at the table is $\widehat{Calories} = 560.7 - 3.08\,Time$. For each additional minute spent at the table, the model predicts that the number of calories consumed will be approximately 3.08 fewer. Only 42.1% of the variability in the number of calories consumed can be accounted for by the variability in time spent at the table. The residuals plot shows no pattern, so the linear model is appropriate, if not terribly useful for prediction.

## R2.39. Tobacco and alcohol.

The first concern about these data is that they consist of averages for regions in Great Britain, not individual households. Any conclusions reached can only be about the regions, not the individual households living there. The second concern is the data point for Northern Ireland. This point has high leverage, since it has the highest household tobacco spending and the lowest household alcohol spending. With this point included, there appears to be only a weak positive association between tobacco and alcohol spending. Without the point, the association is much stronger. In Great Britain, with the exception of Northern Ireland, higher levels of household spending on tobacco are associated with higher levels of household spending on tobacco. It is not necessary to make the linear model, since we have the household averages for the regions in Great Britain, and the model wouldn't be useful for predicting in other countries or for individual households in Great Britain.

## R2.41. Models.

a) $\hat{y} = 2 + 0.8 \ln x$
$\hat{y} = 2 + 0.8 \ln(10)$
$\hat{y} \approx 3.842$

b) $\log \hat{y} = 5 - 0.23x$
$\log \hat{y} = 5 - 0.23(10)$
$\log \hat{y} = 2.7$
$\hat{y} = 10^{2.7} \approx 501.187$

c) $\dfrac{1}{\sqrt{\hat{y}}} = 17.1 - 1.66x$
$\dfrac{1}{\sqrt{\hat{y}}} = 17.1 - 1.66(10) = 0.5$
$\hat{y} = \dfrac{1}{0.5^2} = 4$

**R2.43. Vehicle weights.**

a)

$$\widehat{Wt} = 10.85 + 0.64\,scale$$

$$\widehat{Wt} = 10.85 + 0.64\,(31.2)$$

$$\widehat{Wt} = 30.818$$

According to the model, a truck with a scale weight of 31,200 pounds is expected to weigh 30,818 pounds.

b) If the actual weight of the truck is 32,120 pounds, the residual is 32,120 – 30,818 = 1302 pounds. The model underestimated the weight.

c)

$$\widehat{Wt} = 10.85 + 0.64\,scale$$

$$\widehat{Wt} = 10.85 + 0.64(35.590)$$

$$\widehat{Wt} = 33.6276 \text{ thousand pounds}$$

The predicted weight of the truck is 33,627.6 pounds. If the residual is –2440 pounds, the actual weight of the truck is 33,627.6 – 2440 = 31,187.6 pounds.

d) $R^2 = 93\%$, so the model explains 93% of the variability in weight, but some of the residuals are 1000 pounds or more. If we need to be more accurate than that, then this model will not work well.

e) Negative residuals will be more of a problem. Police would be issuing tickets to trucks whose weights had been overestimated by the model. The U.S. justice system is based upon the principle of innocence until guilt is proven. These truckers would be unfairly ticketed, and that is worse than allowing overweight trucks to pass.

**R2.45. Down the drain.**

The association between diameter of the drain plug and drain time of this water tank is strong, curved, and negative. Tanks with larger drain plugs have lower drain times. The linear model is not appropriate for the curved association, so several re-expressions of the data were tried. The best one was the reciprocal square root re-expression, resulting in the equation

$$\frac{1}{\sqrt{DrainTime}} = 0.00243 + 0.219\,Diameter.$$

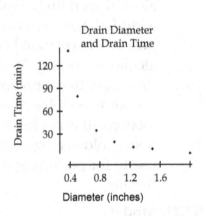

Drain Diameter and Drain Time

The re-expressed data is nearly linear, and although the residuals plot might still indicate some pattern and has one large residual, this is the best of the models examined. The model explains 99.7% of the variability in drain time.

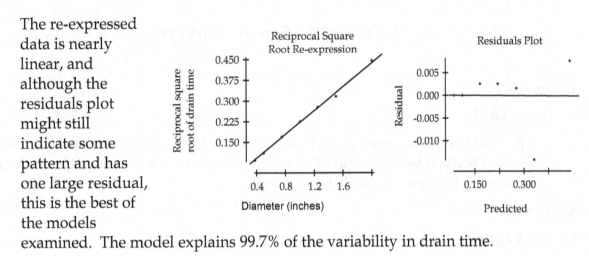

**R2.47. Companies assets and sales.**

a) According to the model, *LogAssets* increase on average by 0.868 Log$ per Log$ of sales for both banks and other companies.

b) Bank assets are, on average, 0.946 Log$ higher than other companies, after allowing for the linear effect of *LogSales*.

c) Yes, the use of the variable *Banks* is appropriate for these data. On the plot, the regression lines are roughly parallel.

**R2.49. Real estate, bathrooms.**

a) After accounting for the linear effect of *Living Area*, houses with another *Bathroom* on average, have a *Price* that is $75,020 higher.

b) The slope of the partial regression line is the *Bathrooms* coefficient, 75,020.3.

**R2.51. Penguins again.**

a) After accounting for the linear effect of *Depth*, the *log(Heart rate)* decreases by 0.045 bpm per minute *Duration* of the dive.

b) Yes, the coefficient of *Duration* is a good estimate of the relationship of *Duration* to (log) *Heart rate*. There is a strong, negative, linear relationship. The longer the *Duration*, the slower the *Heart rate*.

# Chapter 10 – Sample Surveys

**Section 10.1**

## 1. Texas A&M.

The A&M administrators should take a survey. They should sample a part of the student body, selecting respondents with a randomization method. They should be sure to draw a sufficiently large sample.

**Section 10.2**

## 3. A&M again.

The proportion in the sample is a statistic. The proportion of all students is the parameter of interest. The statistic estimates that parameter, but is not likely to be exactly the same.

**Section 10.3**

## 5. Sampling students.

This is not an SRS. Although each student may have an equal chance to be in the survey, groups of friends who choose to sit together will either all be in or out of the sample, so the selection is not independent.

**Section 10.4**

## 7. Sampling A&M students.

a) This is a cluster sample, with each selected dormitory as a cluster.

b) This is a stratified sample, stratified by class year.

c) This systematic sample, with a randomized starting point.

**Section 10.6**

## 9. Survey students.

Several terms are poorly defined. The survey needs to specify the meaning of "family" for this purpose and the meaning of "higher education." The term "seek" may also be poorly defined (for example, would applying to college but not being admitted qualify for seeking more education?)

**Section 10.7**

## 11. Student samples.

a) This would suffer from voluntary response bias. Only those students who saw the advertisement could even be part of the sample, and only those who choose to (and are able to) go to the website would actually be in the sample.

b) This would be a convenience sample, as well as suffer from voluntary response bias.

## Chapter Exercises.

### 13. Roper.

a) Roper is not using a simple random sample. The samples are designed to get 500 males and 500 females. This would be very unlikely to happen in a simple random sample.

b) They are using stratified sample, with two strata, males and females.

### 15. Drug tests.

a) This is a cluster sample, with teams being the clusters.

b) Cluster sampling is a reasonable solution to the problem of randomly sampling players because an entire team can be sampled easily. It would be much more difficult to randomly sample players from many different teams on the same day.

### 17. Medical treatments.

a) **Population** – Unclear, but possibly all U.S. adults.

b) **Parameter** – Proportion who have used and have benefitted from alternative medical treatments.

c) **Sampling Frame** – All Consumers Union subscribers.

d) **Sample** – Those subscribers who responded.

e) **Method** – Not specified, but probably a questionnaire mailed to all subscribers.

f) **Left Out** – Those who are not Consumers Union subscribers.

g) **Bias** – Voluntary response bias. Those who respond may have strong feelings one way or another.

### 19. Mayoral race.

a) **Population** – City voters.

b) **Parameter** – Not clear. They might be interested in the percentage of voters favoring various issues.

c) **Sampling Frame** – All city residents

d) **Sample** – As many residents as they can find in one block from each district. No randomization is specified, but hopefully a block is selected at random within each district.

e) **Method** – Multistage sampling, stratified by district and clustered by block.

f) **Left Out** – People not home during the time of the survey.

g) **Bias** – Convenience sampling. Once the block is randomly chosen as the cluster, every resident living in that block should be surveyed, not just those that were conveniently available. A random sample of each block could be also be taken, but we wouldn't refer to that as "cluster" sampling, but rather multi-stage, with stratification by district, a simple random sample of one block within each district, and another simple random sample of residents within the block.

**21. Roadblock.**

a) **Population** – All cars.

b) **Parameter** – Proportion of cars with up-to-date (or out-of-date) registrations, insurance, or safety inspections.

c) **Sampling Frame** – Cars on that road.

d) **Sample** – Cars stopped by the roadblock.

e) **Method** – Cluster sample of an area, stopping all cars within the cluster.

f) **Left Out** – Drivers that did not take that road, or traveled that road at a different time.

g) **Bias** – Undercoverage. The cars stopped might not be representative of all cars because of time of day and location. The locations are probably not chosen randomly, so might represent areas in which it is easy to set up a roadblock, resulting in a convenience sample.

**23. Milk samples.**

a) **Population** – Dairy farms.

b) **Parameter** – Whether or not the milk contains dirt, antibiotics, or other foreign matter.

c) **Sampling Frame** – All dairy farms

d) **Sample** – Not specified, but probably a random sample of farms.

e) **Method** – not specified

f) **Left Out** – Nothing.

g) **Bias** – Unbiased, as long as the day of inspection is randomly chosen. This might not be the case, however, since the farms might be spread out over a wide geographic area. Inspectors might tend to visit farms that are near one another on the same day, resulting in a convenience sample.

**25. Another mistaken poll.**

The newspaper's faulty prediction was more likely to be due to sampling error. The description of the sampling method suggests that samples should be representative of the voting population. Random chance in selecting the individuals who were polled means that sample statistics will vary from the population parameter, perhaps by quite a bit.

**27. Parent opinion, part 2.**

a) This sampling method suffers from voluntary response bias. Only those who see the show and feel strongly will call.

b) Although this method may result in a more representative sample than the method in part **a)**, this is still a voluntary response sample. Only strongly motivated parents attend PTA meetings.

c) This is multistage sampling, stratified by elementary school and then clustered by grade. This is a good design, as long as the parents in the class respond. There should be follow-up to get the opinions of parents who do not respond.

d) This is systematic sampling. As long as a starting point is randomized, this method should produce reliable data.

**29. Playground.**

The managers will only get responses from people who come to the park to use the playground. Parents who are dissatisfied with the playground may not come.

**31. Playground, act two.**

The first sentence points our problems that the respondent may not have noticed, and might lead them to feel they should agree. The last phrase mentions higher fees, which could make people reject improvements to the playground.

**33. Banning ephedra.**

a) This is a voluntary response survey. The large sample will still be affected by any biases in the group of people that choose to respond.

b) The wording seems fair enough. It states the facts, and gives voice to both sides of the issue.

c) The sampling frame is, at best, those who visit this particular site, and even then depends of their volunteering to respond to the question.

d) This statement is true.

**35. More survey questions.**

a) The question seems unbiased.

**b)** The question is biased toward "yes" because of the phrase "great tradition". A better question: "Do you favor continued funding for the space program?"

## 37. Cell phone survey.

Cell phones are more likely to be used by younger individuals. This will result in an undercoverage bias. As cell phone use grows, this will be less of a problem. Also, many cell phone plans require the users to pay airtime for incoming calls. That seems like a sure way to irritate the respondent, and result in response bias toward negative responses.

## 39. Fuel economy.

**a)** The statistic calculated is the mean mileage for the last six fill-ups.

**b)** The parameter of interest is the mean mileage for the vehicle.

**c)** The driving conditions for the last six fill-ups might not be typical of the overall driving conditions. For instance, the last six fill-ups might all be in winter, when mileage might be lower than expected.

**d)** The EPA is trying to estimate the mean gas mileage for all cars of this make, model, and year.

## 41. Happy workers?

**a)** A small sample will probably consist mostly laborers, with few supervisors, and maybe no project managers. Also, there is a potential for response bias based on the interviewer if a member of management asks directly about discontent. Workers who want to keep their jobs will likely tell the management that everything is fine!

**b)** Assign a number from 001 to 439 to each employee. Use a random number table or software to select the sample.

**c)** The simple random sample might not give a good cross section of the different types of employees. There are relatively few supervisors and project managers, and we want to make sure their opinions are noted, as well as the opinions of the laborers.

**d)** A better strategy would be to stratify the sample by job type. Sample a certain percentage of each job type.

**e)** Answers will vary. Assign each person a number from 01-14, and generate 2 usable random numbers from a random number table or software.

**43. A fish story.**

What conclusions they may be able to make will depend on whether fish with discolored scales are equally likely to be caught as those without. It also depends on the level of compliance by fisherman. If fish are not equally likely to be caught, or fishermen more disposed to bring discolored fish, the results will be biased.

**45. Sampling methods.**

a) This method would probably result in undercoverage of those doctors that are not listed in the Yellow Pages. Using the "line listings" seems fair, as long as all doctors are listed, but using the advertisements would not be a typical list of doctors.

b) This method is not appropriate. This cluster sample will probably contain listings for only one or two types of businesses, not a representative cross-section of businesses.

# Chapter 11 – Experiments and Observational Studies

**Section 11.1**

**1. Steroids.**

This is an observational study because the sports writer is not randomly assigning players to take steroids or not take steroids; the writer is merely observing home run totals between two eras. It would be unwise to conclude steroids caused any increases in home runs because we need to consider other factors besides steroids — factors possibly leading to more home runs include better equipment, players training more in the offseason, smaller ballparks, better scouting techniques, etc.

**Section 11.2**

**3. Tips.**

Each of the 40 deliveries is an experimental unit. He has randomized the experiment by flipping a coin to decide whether or not to phone.

**5. Tips II.**

The factor is calling, and the levels are whether or not he calls the customer. The response variable is the tip percentage for each delivery.

**Section 11.3**

**7. Tips again.**

By calling some customers but not others during the same run, the driver has controlled many variables, such as day of the week, season, and weather. The experiment was randomized because he flipped a coin to determine whether or not to phone and it was replicated because he did this for 40 deliveries.

**Section 11.4**

**9. More tips.**

Because customers don't know about the experiment, those that are called don't know that others are not, and vice versa. Thus, the customers are blind. That would make this a single-blind study. It can't be double-blind because the delivery driver must know whether or not he phones.

**Section 11.5**

**11. Block that tip.**

Yes. Driver is now a block. The experiment is randomized within each block. This is a good idea because some drivers might generally get higher tips than others, but the goal of the experiment is to study the effect of phone calls. Blocking on driver eliminates the variability in tips inherent to the driver.

**Section 11.6**

**13. Confounded tips.**

Answers may vary. The cost or size of a delivery may confound his results. Larger orders may generally tip a higher or lower percentage of the bill.

**Chapter Exercises**

**15. Standardized test scores.**

a) No, this is not an experiment. There are no imposed treatments. This is a retrospective observational study.

b) We cannot conclude that the differences in score are caused by differences in parental income. There may be lurking variables that are associated with both SAT score and parental income.

**17. MS and vitamin D.**

a) This is a retrospective observational study.

b) This is an appropriate choice, since MS is a relatively rare disease.

c) The subjects were U.S. military personnel, some of whom had developed MS.

d) The variables were the vitamin D blood levels and whether or not the subject developed MS.

**19. Menopause.**

a) This was a randomized, comparative, placebo-controlled experiment.

b) Yes, such an experiment is the right way to determine whether black cohosh is an effective treatment for hot flashes.

c) The subjects were 351 women, aged 45 to 55 who reported at least two hot flashes a day.

d) The treatments were black cohosh, a multi-herb supplement, plus advice to consume more soy foods, estrogen, and a placebo. The response was the women's self-reported symptoms, presumably the frequency of hot flashes.

**21. a)** This is an experiment, since treatments were imposed.
  **b)** The subjects studied were 30 patients with bipolar disorder.
  **c)** The experiment has 1 factor (omega-3 fats from fish oil), at 2 levels (high dose of omega-3 fats from fish oil and no omega-3 fats from fish oil).
  **d)** 1 factor, at 2 levels gives a total of 2 treatments.
  **e)** The response variable is "improvement", but there is no indication of how the response variable was measured.
  **f)** There is no information about the design of the experiment.

**g)** The experiment is blinded, since the use of a placebo keeps the patients from knowing whether or not they received the omega-3 fats from fish oils. It is not stated whether or not the evaluators of the "improvement" were blind to the treatment, which would make the experiment double-blind.

**h)** Although it needs to be replicated, the experiment can determine whether or not omega-3 fats from fish oils cause improvements in patients with bipolar disorder, at least over the short term. The experiment design would be stronger is it were double-blind.

**23. a)** This is an observational study. The researchers are simply studying traits that already exist in the subjects, not imposing new treatments.

**b)** This is a prospective study. The subjects were identified first, then traits were observed.

**c)** The subjects are roughly 200 men and women with moderately high blood pressure and normal blood pressure. There is no information about the selection method.

**d)** The parameters of interest are difference in memory and reaction time scores between those with normal blood pressure and moderately high blood pressure.

**e)** An observational study has no random assignment, so there is no way to know that high blood pressure caused subjects to do worse on memory and reaction time tests. A lurking variable, such as age or overall health, might have been the cause. The most we can say is that there was an association between blood pressure and scores on memory and reaction time tests in this group, and recommend a controlled experiment to attempt to determine whether or not there is a cause-and-effect relationship.

**25. a)** This is an experiment, since treatments were imposed on randomly assigned groups.

**b)** 24 post-menopausal women were the subjects in this experiment.

**c)** There is 1 factor (type of drink), at 2 levels (alcoholic and non-alcoholic). (Supplemental estrogen is not a factor in the experiment, but rather a blocking variable. The subjects were not given estrogen supplements as part of the experiment.)

**d)** 1 factor, with 2 levels, is 2 treatments.

**e)** The response variable is an increase in estrogen level.

**f)** This experiment utilizes a blocked design. The subjects were blocked by whether or not they used supplemental estrogen. This design reduces variability in the response variable of estrogen level that may be associated with the use of supplemental estrogen.

**g)** This experiment does not use blinding.

**h)** This experiment indicates that drinking alcohol leads to increased estrogen level among those taking estrogen supplements.

**27. a)** This is an observational study.

   **b)** The study is retrospective. Results were obtained from pre-existing church records.

   **c)** The subjects of the study are women in Finland. The data were collected from church records dating 1640 to 1870, but the selection process is unknown.

   **d)** The parameter of interest is difference in average lifespan between mothers of sons and daughters.

   **e)** For this group, having sons was associated with a decrease in lifespan of an average of 34 weeks per son, while having daughters was associated with an unspecified increase in lifespan. As there is no random assignment, there is no way to know that having sons caused a decrease in lifespan.

**29. a)** This is an observational study. (Although some might say that the sad movie was "imposed" on the subjects, this was merely a stimulus used to trigger a reaction, not a treatment designed to attempt to influence some response variable. Researchers merely wanted to observe the behavior of two different groups when each was presented with the stimulus.)

   **b)** The study is prospective. Researchers identified subjects, and then observed them after the sad movie.

   **c)** The subjects in this study were people with and without depression. The selection process is not stated.

   **d)** The parameter of interest is the difference in crying response between depressed and nondepressed people exposed to sad situations.

   **e)** There is no apparent difference in crying response to sad movies for the depressed and nondepressed groups.

**31. a)** This is an experiment. Subjects were randomly assigned to treatments.

   **b)** The subjects were people experiencing migraines.

   **c)** There are 2 factors (pain reliever and water temperature). The pain reliever factor has 2 levels (pain reliever or placebo), and the water temperature factor has 2 levels (ice water and regular water).

   **d)** 2 factors, at 2 levels each, results in 4 treatments.

   **e)** The response variable is the level of pain relief.

   **f)** The experiment is completely randomized.

   **g)** The subjects are blinded to the pain reliever factor through the use of a placebo. The subjects are not blinded to the water factor. They will know whether they are drinking ice water or regular water.

   **h)** The experiment may indicate whether pain reliever alone or in combination with ice water give pain relief, but patients are not blinded to ice water, so the placebo effect may also be the cause of any relief seen due to ice water.

**33. a)** This is an experiment. Athletes were randomly assigned to one of two exercise programs.
   **b)** The subjects are athletes suffering hamstring injuries.
   **c)** There is one factor (type of exercise), at 2 levels (static stretching, and agility and trunk stabilization).
   **d)** 1 factor, at 2 levels, results in 2 treatments.
   **e)** The response variable is the time before the athletes were able to return to sports.
   **f)** The experiment is completely randomized.
   **g)** The experiment employs no blinding. The subjects know what kind of exercise they do.
   **h)** Assuming that the athletes actually followed the exercise program, this experiment can help determine which of the two exercise programs is more effective at rehabilitating hamstring injuries.

**35. Omega-3.**

The experimenters need to compare omega-3 results to something. Perhaps bipolarity is seasonal and would have improved during the experiment anyway.

**37. Omega-3 revisited.**

   **a)** Subjects' responses might be related to other factors, like diet, exercise, or genetics. Randomization should equalize the two groups with respect to unknown factors.

   **b)** More subjects would minimize the impact of individual variability in the responses, but the experiment would become more costly and time-consuming.

**39. Omega-3 finis.**

The researchers believe that people who engage in regular exercise might respond differently to the omega-3. This additional variability could obscure the effectiveness of the treatment.

**41. Injuries.**

Answers may vary. Use a random-number generator to randomly select 24 numbers from 01 to 24 without replication. Assign the first 8 numbers to the first group, the second 8 numbers to the second group, and the third 8 numbers to the third group. If an athlete states that he would prefer the other program, he should not be allowed to switch, since a successful experiment requires randomization. In fact, he should probably be eliminated from the study, since there is a risk that he will not follow the directions for his assigned program. He may instead opt to follow the protocols for the other program anyway, since he has an obvious preference. This would invalidate the results of the experiment.

**43. Shoes.**

**a)** First, the manufacturers are using athletes who have a vested interest in the success of the shoe by virtue of their sponsorship. They should try to find some volunteers that aren't employed by the company! Second, they should randomize the order of the runs, not run all the races with the new shoes second. They should blind the athletes by disguising the shoes, if possible, so they don't know which is which. The experiment could be double blinded, as well, by making sure that the timers don't know which shoes are being tested at any given time. Finally, they should replicate several times since times will vary under both shoe conditions.

**b)** First of all, the problems identified in part a would have to be remedied before *any* conclusions can be reached. Even if this is the case, the results cannot be generalized to all runners. This experiment compares effects of the shoes on speed for Olympic class runners, not runners in general.

**45. Hamstrings.**

**a)** Allowing the athletes to choose their own treatments could confound the results. Other issues such as severity of injury, diet, age, etc., could also affect time to heal, and randomization should equalize the two treatment groups with respect to any such variables.

**b)** A control group could have revealed whether either exercise program was better (or worse) than just letting the injury heal without exercise.

**c)** Although the athletes cannot be blinded, the doctors who approve their return to sports should not know which treatment the subject had engaged in.

**d)** It's difficult to say with any certainty, since we aren't sure if the distributions of return times are unimodal and roughly symmetric, and contain no outliers. Otherwise, the use of mean and standard deviation as measures of center and spread is questionable. Assuming mean and standard deviation are appropriate measures, the subjects who exercised with agility and trunk stabilization had a mean return time of 22.2 days compared to the static stretching group, with a mean return time of 37.4 days. The agility and trunk stabilization group also had a much more consistent distribution of return times, with a standard deviation of 8.3 days, compared to the standard deviation of 27.6 days for the static stretching group. This appears to be a statistically significant difference.

**47. Mozart.**

**a)** The differences in spatial reasoning scores between the students listening to Mozart and the students sitting quietly were more than would have been expected from ordinary sampling variation.

**b)**

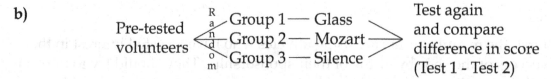

Pre-tested volunteers — R a n d o m — Group 1 — Glass / Group 2 — Mozart / Group 3 — Silence → Test again and compare difference in score (Test 1 - Test 2)

**c)** The Mozart group seems to have the smallest median difference in spatial reasoning test score and thus the *least* improvement, but there does not appear to be a significant difference.

**d)** No, the results do not prove that listening to Mozart is beneficial. If anything, there was generally less improvement. The difference does not seem significant compared with the usual variation one would expect between the three groups. Even if type of music has no effect on test score, we would expect some variation between the groups.

**49. Wine.**

**a)** This is a prospective observational study. The researchers followed a group of children born at a Copenhagen hospital between 1959 and 1961.

**b)** The results of the Danish study report a link between high socioeconomic status, education, and wine drinking. Since people with high levels of education and higher socioeconomic status are also more likely to be healthy, the relation between health and wine consumption might be explained by the confounding variables of socioeconomic status and education.

**c)** Studies such as these prove none of these. While the variables have a relation, there is no indication of a cause-and-effect relationship. The only way to determine causation is through a controlled, randomized, and replicated experiment.

**51. Dowsing.**

**a)** Arrange the 20 containers in 20 separate locations. Number the containers 01 – 20, and use a random number generator to identify the 10 containers that should be filled with water.

**b)** We would expect the dowser to be correct about 50% of the time, just by guessing. A record of 60% (12 out of 20) does not appear to be significantly different than the 10 out of 20 expected.

**c)** Answers may vary. A high level of success would need to be observed. 90% to 100% success (18 to 20 correct identifications) would be convincing.

## 53. Reading.

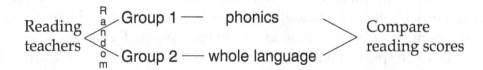

Answers may vary. This experiment has 1 factor (reading program), at 2 levels (phonics and whole language), resulting in 2 treatments. The response variable is reading score on an appropriate reading test after a year in the program. After randomly assigning students to teachers, randomly assign half the reading teachers in the district to use each method. There may be variation in reading score based on school within the district, as well as by grade. Blocking by both school and grade will reduce this variation.

## 55. Weekend deaths.

**a)** The difference between death rate on the weekend and death rate during the week is greater than would be expected due to natural sampling variation.

**b)** This was a prospective observational study. The researchers identified hospitals in Ontario, Canada, and tracked admissions to the emergency rooms. This certainly cannot be an experiment. People can't be assigned to become injured on a specific day of the week!

**c)** Waiting until Monday, if you were ill on Saturday, would be foolish. There are likely to be confounding variables that account for the higher death rate on the weekends. For example, people might be more likely to engage in risky behavior on the weekend.

**d)** Alcohol use might have something to do with the higher death rate on the weekends. Perhaps more people drink alcohol on weekends, which may lead to more traffic accidents, and higher rates of violence during these days of the week.

## 57. Beetles.

Answers may vary. This experiment has 1 factor (pesticide), at 3 levels (pesticide A, pesticide B, no pesticide), resulting in 3 treatments. The response variable is the number of beetle larvae found on each plant. Randomly select a third of the plots to be sprayed with pesticide A, a third with pesticide B, and a third to be sprayed with no pesticide (since the researcher also wants to know whether the pesticides even work at all). To control the experiment, the plots of land should be as similar as possible, with regard to amount of sunlight, water, proximity to other plants, etc. If not, plots with similar characteristics should be blocked together. If possible, use some inert substance as a placebo pesticide on the control group, and do not tell the counters of the beetle larvae which plants have been treated with pesticides. After a given period of time, count the number of beetle larvae on each plant and compare the results.

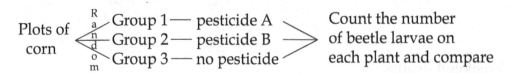

## 59. Safety switch.

Answers may vary. This experiment has 1 factor (hand), at 2 levels (right, left), resulting in 2 treatments. The response variable is the difference in deactivation time between left and right hand. Find a group of volunteers. Using a matched design, we will require each volunteer to deactivate the machine with his or her left hand, as well as with his or her right hand. Randomly assign the left or right hand to be used first. Hopefully, this will equalize any variability in time that may result from experience gained after deactivating the machine the first time. Complete the first attempt for the whole group. Now repeat the experiment with the alternate hand. Check the differences in time for the left and right hands. Since the response variable is difference in times for each hand, workers should be blocked into groups based on their dominant hand. Another way to account for this difference would be to use the absolute value of the difference as the response variable. We are interested in whether or not the difference is significantly different from the zero difference we would expect if the machine were just as easy to operate with either hand.

## 61. Skydiving, anyone?

a) There is 1 factor, jumping, with 2 levels, with and without a working parachute.

b) You would need some (dim-witted) volunteers skydivers as the subjects.

c) A parachute that looked real, but didn't open, would serve as the placebo.

d) 1 factor at 2 levels is 2 treatments, a good parachute and a placebo parachute.

e) The response variable is whether the skydiver survives the jump (or the extent of injuries).

f) All skydivers should jump from the same altitude, in similar weather conditions, and land on similar surfaces.

g) Make sure that you randomly assign the skydivers to the parachutes.

h) The skydivers (and the distributers of the parachutes) shouldn't know who got a working chute. Additionally, the people evaluating the subjects after the jumps should not be told who had a real chute, either.

# Review of Part III – Gathering Data

**R3.1.** The researchers performed a prospective observational study, since the children were identified at birth and examined at ages 8 and 20. There were indications of behavioral differences between the group of "preemies", and the group of full-term babies. The "preemies" were less likely to engage in risky behaviors, like use of drugs and alcohol, teen pregnancy, and conviction of crimes. This may point to a link between premature birth and behavior, but there may be lurking variables involved. Without a controlled, randomized, and replicated experiment, a cause-and-effect relationship cannot be determined.

**R3.3.** The researchers at the Purina Pet Institute performed an experiment, matched by gender and weight. The experiment had one factor (diet), at two levels (allowing the dogs to eat as much as they want, or restricted diet), resulting in two treatments. One of each pair of similar puppies was randomly assigned to each treatment. The response variable was length of life. The researchers were able to conclude that, on average, dogs with a lower-calorie diet live longer.

**R3.5.** This is a completely randomized experiment, with the treatment being receiving folic acid or not (one factor, two levels). Treatments were assigned randomly and the response variable is the number of precancerous growths, or simply the occurrence of additional precancerous growths. Neither blocking nor matching is mentioned, but in a study such as this one, it is likely that researchers and patients are blinded. Since treatments were randomized, it seems reasonable to generalize results to all people with precancerous polyps, though caution is warranted since these results contradict a previous study.

**R3.7.** The fireworks manufacturers are sampling. No information is given about the sampling procedure, so hopefully the tested fireworks are selected randomly. It would probably be a good idea to test a few of each type of firework, so stratification by type seems likely. The population is all fireworks produced each day, and the parameter of interest is the proportion of duds. With a random sample, the manufacturers can make inferences about the proportion of duds in the entire day's production, and use this information to decide whether or not the day's production is suitable for sale.

**R3.9.** This is an observational retrospective study. Researcher can conclude that for anyone's lunch, even when packed with ice, food temperatures are rising to unsafe levels.

**R3.11.** This is an experiment, with a control group being the genetically engineered mice who received no antidepressant and the treatment group being the mice who received the drug. The response variable is the amount of plaque in their brains after one dose and after four months. There is no mention of blinding or matching. Conclusions can be drawn to the general population of mice and we should assume treatments were randomized. To conclude the same for humans would be risky, but researchers might propose an experiment on humans based on this study.

**R3.13.** The researchers performed an experiment. There is one factor (gene therapy), at two levels (gene therapy and no gene therapy), resulting in two treatments. The experiment is completely randomized. The response variable is heart muscle condition. The researchers can conclude that gene therapy is responsible for stabilizing heart muscle in laboratory rats.

**R3.15.** The orange juice plant depends on sampling to ensure the oranges are suitable for juice. The population is all of the oranges on the truck, and the parameter of interest is the proportion of unsuitable oranges. The procedure used is a random sample, stratified by location in the truck. Using this well-chosen sample, the workers at the plant can estimate the proportion of unsuitable oranges on the truck, and decide whether or not to accept the load.

**R3.17.** Observational retrospective study, performed as a telephone-based randomized survey. Based on the excerpt, it seems reasonable to conclude that more education is associated with a higher Emotional Health Index score, but to insist on causality would be faulty reasoning.

**R3.19. Commuter sample.**

The sample shows skewness to the right that is in the population. The sample suggests there may be high outliers, but the population has a smooth tail and no outliers.

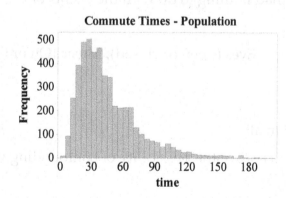

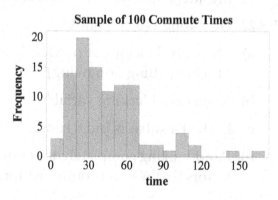

**R3.21. Alternate day fasting.**

a) 100 obese persons.

b) Restrictive diet, alternate fasting diet, control (no special diet).

c) Weight loss (reported as % of initial weight).

d) No, it could not be blinded because participants must know what they are eating.

e) No, it is not necessary that participants be a random sample from the population. All that is needed is that they be randomly assigned to treatment groups.

**R3.23. Tips.**

a) The waiters performed an experiment, since treatments were imposed on randomly assigned groups. This experiment has one factor (candy), at two levels (candy or no candy), resulting in two treatments. The response variable is the percentage of the bill given as a tip.

b) If the decision whether to give candy or not was made before the people were served, the server may have subconsciously introduced bias by treating the customers better. If the decision was made just before the check was delivered, then it is reasonable to conclude that the candy was the cause of the increase in the percentage of the bill given as a tip.

c) "Statistically significant" means that the difference in the percentage of tips between the candy and no candy groups was more than expected due to sampling variability.

**R3.25. Timing.**

There will be voluntary response bias, and results will mimic those only of the visitors to Sodahead.com and not the general U.S. population. The question is leading responders to answer "yes" though many might understand that the president's timing for his vacation had nothing to do with the events of the week.

**R3.27. How long is 30 seconds?**

a) There are 3 factors, at two levels each. Eyes (open or closed), Music (On or Off), Moving (sitting or moving).

b) *Subject* is a blocking variable.

c) Each of 4 subjects did 8 runs, so 32 in all.

d) Randomizing completely is better to reduce the possibility of confounding with factors that weren't controlled for.

**R3.29. Homecoming.**

**a)** Since telephone numbers were generated randomly, every number that could possibly occur in that community had an equal chance of being selected. This method is "better" than using the phone book, because unlisted numbers are also possible. Those community members who deliberately do not list their phone numbers might not consider this method "better"!

**b)** Although this method results in a simple random sample of phone numbers, it does not result in a simple random sample of residences. Residences without a phone are excluded, and residences with more than one phone have a greater chance of being included.

**c)** No, this is not a SRS of local voters. People who respond to the survey may be of the desired age, but not registered to vote. Additionally, some voters who are contacted may choose not to participate.

**d)** This method does not guarantee an unbiased sample of households. Households in which someone answered the phone may be more likely to have someone at home when the phone call was generated. The attitude about homecoming of these households might not be the same as the attitudes of the community at large.

**R3.31. Smoking and Alzheimer's.**

**a)** The studies do not prove that smoking offers any protection from Alzheimer's. The studies merely indicate an association. There may be other variables that can account for this association.

**b)** Alzheimer's usually shows up late in life. Since smoking is known to be harmful, perhaps smokers have died of other causes before Alzheimer's can be seen.

**c)** The only way to establish a cause-and-effect relationship between smoking and Alzheimer's is to perform a controlled, randomized, and replicated experiment. This is unlikely to ever happen, since the factor being studied, smoking, has already been proven harmful. It would be unethical to impose this treatment on people for the purposes of this experiment. A prospective observational study could be designed in which groups of smokers and nonsmokers are followed for many years and the incidence of Alzheimer's disease is tracked.

### R3.33. Sex and violence.

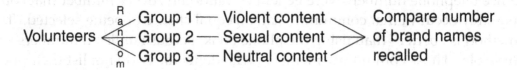

This experiment has one factor (program content), at three levels (violent, sexual, and neutral), resulting in three treatments. The response variable is the number of brand names recalled after watching the program. Numerous subjects will be randomly assigned to see shows with violent, sexual, or neutral content. They will see the same commercials. After the show, they will be interviewed for their recall of brand names in the commercials.

### R3.35. Age and party 2008.

a)  The number of respondents is roughly the same for each age category. This may indicate a sample stratified by age category, although it may be a simple random sample.

b)  1530 Democrats were surveyed. $\frac{1530}{4002} \approx 38.2\%$ of the people surveyed were Democrats.

c)  We don't know. If data were collected from voting precincts that are primarily Democratic or primarily Republican, that would bias the results. Because the survey was commissioned by NBC News, we can assume the data collected are probably reliable.

d)  The pollsters were probably attempting to determine whether or not political party is associated with age.

### R3.37. Save the grapes.

This experiment has one factor (bird control device), at three levels (scarecrow, netting, and no device), resulting in three treatments. Randomly assign different plots in the vineyard to the different treatments, making sure to ensure adequate separation of plots, so that the possible effect of the scarecrow will not be confounded with the other treatments. The response variable to be measured at the end of the season is the proportion of bird-damaged grapes in each plot.

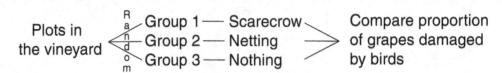

### R3.39. Acupuncture.

a) The "fake" acupuncture was the control group. In an experiment, all subjects must be treated as alike as possible. If there were no "fake" acupuncture, subjects would know that they had not received acupuncture, and might react differently. Of course, all volunteers for the experiment must be aware of the possibility of being randomly assigned to the control group.

b) Experiments always use volunteers. This is not a problem, since experiments are testing response to a treatment, not attempting to determine an unknown population parameter. The randomization in an experiment is random assignment to treatment groups, not random selection from a population. Voluntary response is a problem when sampling, but is not an issue in experimentation. In this case, it is probably reasonable to assume that the volunteers have similar characteristics to others in the population of people with chronic lower back pain.

c) There were differences in the amount of pain relief experienced by the two groups, and these differences were large enough that they could not be explained by natural variation alone. Researchers concluded that both proper and "fake" acupuncture reduced back pain.

### R3.41. Security.

a) To ensure that passengers from first-class, as well as coach, get searched, select 2 passengers from first-class and 12 from coach. Using this stratified random sample, 10% of the first-class passengers are searched, as are 10% of the coach passengers.

b) Answers will vary. Number the passengers alphabetically, with 2-digit numbers. Bergman = 01, Bowman = 02, and so on, ending with Testut = 20. Read the random digits in pairs, ignoring pairs 21 to 99 and 00, and ignoring repeated pairs.

```
65|43|67|11|        27|04|          The passengers selected for search
XX XX XX Fontana    XX Castillo      from first-class are Fontana and
                                     Castillo.
```

c) Number the passengers alphabetically, with 3 digit numbers, 001 to 120. Use the random number table to generate 3-digit numbers, ignoring numbers 121 to 999 and 000, and ignoring repeated numbers. Search the passengers corresponding to the first 12 valid numbers generated.

# Chapter 12 – From Randomness to Probability

## Section 12.1

**1. Flipping a coin.**

In the long run, a fair coin will generate 50% heads and 50% tails, approximately. But for each flip we cannot predict the outcome.

**3. Flipping a coin II.**

There is no law of averages for the short run. The first five flips do not affect the sixth flip.

## Section 12.2

**5. Wardrobe.**

a) There are a total of 10 shirts, and 3 of them are red. The probability of randomly selecting a red shirt is $3/10 = 0.30$.

b) There are a total of 10 shirts, and 8 of them are not black. The probability of randomly selecting a shirt that is not black is $8/10 = 0.80$.

## Section 12.3

**7. Cell phones and surveys.**

a) If 51% of homes don't have a landline, then 49% of them do have a landline. The probability that all 5 houses have a landline is $(0.49)^5 \approx 0.028$.

b) $P(\text{at least one without landline}) = 1 - P(\text{all landlines}) = 1 - (0.49)^5 \approx 0.972$.

c) $P(\text{at least one with landline}) = 1 - P(\text{no landlines}) = 1 - (0.51)^5 \approx 0.965$

**9. Pet ownership.**

$$P(\text{dog or cat}) = P(\text{dog}) + P(\text{cat}) - P(\text{dog and cat})$$
$$= 0.25 + 0.29 - 0.12 = 0.42$$

**Section 12.4**

**11. Sports.**

|  | Football | No Football | Total |
|---|---|---|---|
| Basketball | 27 | 13 | 40 |
| No Basketball | 38 | 22 | 60 |
| Total | 65 | 35 | 100 |

$$P(\text{football} \mid \text{basketball}) = \frac{P(\text{football and basketball})}{P(\text{basketball})} = \frac{\frac{27}{100}}{\frac{40}{100}} = 0.675$$

(Or, use the table. Of the 40 people who like to watch basketball, 27 people also like to watch football. 27/40 = 0.675)

**13. Late to the train.**

$$P(\text{let out late and missing train}) = P(\text{let out late}) \times P(\text{missing train} \mid \text{let out late})$$
$$= (0.30)(0.45) = 0.135$$

**Section 12.5**

**15. Titanic.**

The overall survival rate, $P(S)$, was 0.323, yet the survival rate for first class passengers, $P(S \mid FC)$, was 0.625. Since, $P(S) \neq P(S \mid FC)$, survival and ticket class are not independent. Rather, survival rate depended on class.

**Section 12.6**

**17. Facebook.**

|  | US | Not US | Total |
|---|---|---|---|
| Log on Every Day | 0.20 | 0.30 | 0.50 |
| Do Not Log on Every Day | 0.10 | 0.40 | 0.50 |
| Total | 0.30 | 0.70 | 1.00 |

We have joint probabilities and marginal probabilities, not conditional probabilities, so a table is the better choice.

## 19. Facebook again.

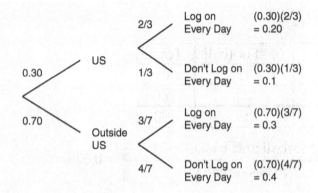

A tree is better because we have conditional and marginal probabilities. The joint probabilities are found at the end of the branches.

### Section 12.7

## 21. Facebook final.

$$P(\text{US} \mid \text{Log on every day}) = \frac{P(\text{US and Log on every day})}{P(\text{Log on every day})} = \frac{0.20}{0.20 + 0.30} = 0.40$$

Knowing that a person logs on every day increases probability that the person is from the United States.

### Chapter Exercises.

## 23. Sample spaces.

a) S = { HH, HT, TH, TT} All of the outcomes are equally likely to occur.

b) S = { 0, 1, 2, 3} All outcomes are not equally likely. A family of 3 is more likely to have, for example, 2 boys than 3 boys. There are three equally likely outcomes that result in 2 boys (BBG, BGB, and GBB), and only one that results in 3 boys (BBB).

c) S = { H, TH, TTH, TTT} All outcomes are not equally likely. For example the probability of getting heads on the first try is $\frac{1}{2}$. The probability of getting three tails is $\left(\frac{1}{2}\right)^3 = \frac{1}{8}$.

d) S = {1, 2, 3, 4, 5, 6} All outcomes are not equally likely. Since you are recording only the larger number (or the number if there is a tie) of two dice, 6 will be the larger when the other die reads 1, 2, 3, 4, or 5. The outcome 2 will only occur when the other die shows 1 or 2.

## 25. Roulette.

If a roulette wheel is to be considered truly random, then each outcome is equally likely to occur, and knowing one outcome will not affect the probability of the next. Additionally, there is an implication that the outcome is not determined through the use of an electronic random number generator.

## 27. Winter.

Although acknowledging that there is no law of averages, Knox attempts to use the law of averages to predict the severity of the winter. Some winters are harsh and some are mild over the long run, and knowledge of this can help us to develop a long-term probability of having a harsh winter. However, probability does not compensate for odd occurrences in the short term. Suppose that the probability of having a harsh winter is 30%. Even if there are several mild winters in a row, the probability of having a harsh winter is still 30%.

## 29. Auto insurance.

a) It would be foolish to insure your neighbor against automobile accidents for $1500. Although you might simply collect $1500, there is a good chance you could end up paying much more than $1500. That risk is not worth the $1500.

b) The insurance company insures many people. The overwhelming majority of customers pay the insurance and never have a claim, or have claims that are lower than the cost of their payments. The few customers who do have a claim are offset by the many who simply send their premiums without a claim. The relative risk to the insurance company is low.

## 31. Spinner.

a) This is a legitimate probability assignment. Each outcome has probability between 0 and 1, inclusive, and the sum of the probabilities is 1.

b) This is a legitimate probability assignment. Each outcome has probability between 0 and 1, inclusive, and the sum of the probabilities is 1.

c) This is not a legitimate probability assignment. Each outcome has probability between 0 and 1, inclusive, but the sum of the probabilities is greater than 1.

d) This is a legitimate probability assignment. Each outcome has probability between 0 and 1, inclusive, and the sum of the probabilities is 1. However, this game is not very exciting!

e) This probability assignment is not legitimate. The sum of the probabilities is 0, and there is one probability, –1.5 , that is not between 0 and 1, inclusive.

## 33. Electronics.

A family may have both a computer and an HDTV. The events are not disjoint, so the Addition Rule does not apply.

**35. Speeders.**

When cars are traveling close together, their speeds are not independent. For example, a car following directly behind another can't be going faster than the car ahead. Since the speeds are not independent, the Multiplication Rule does not apply.

**37. College admissions.**

a) Jorge had multiplied the probabilities.

b) Jorge assumes that being accepted to the colleges are independent events.

c) No. Colleges use similar criteria for acceptance, so the decisions are not independent. Students that meet these criteria are more likely to be accepted at all of the colleges. Since the decisions are not independent, the probabilities cannot be multiplied together.

**39. Car repairs.**

Since all of the events listed are disjoint, the addition rule can be used.

a) $P$(no repairs) $= 1 - P$(some repairs) $= 1 - (0.17 + 0.07 + 0.04) = 1 - (0.28) = 0.72$

b) $P$(no more than one repair) $= P$(no repairs or one repair) $= 0.72 + 0.17 = 0.89$

c) $P$(some repairs) $= P$(one or two or three or more repairs)
$$= 0.17 + 0.07 + 0.04 = 0.28$$

**41. More repairs.**

Assuming that repairs on the two cars are independent from one another, the multiplication rule can be used. Use the probabilities of events from Exercise 27 in the calculations.

a) $P$(neither will need repair) $= (0.72)(0.72) = 0.5184$

b) $P$(both will need repair) $= (0.28)(0.28) = 0.0784$

c) $P$(at least one will need repair) $= 1 - P$ (neither will need repair)
$$= 1 - (0.72)(0.72) = 0.4816$$

**43. Repairs, again.**

a) The repair needs for the two cars must be independent of one another.

b) This may not be reasonable. An owner may treat the two cars similarly, taking good (or poor) care of both. This may decrease (or increase) the likelihood that each needs to be repaired.

**45. Polling.**

a) $P$(household is contacted and household refuses to cooperate)
$= P$(household is contacted)$P$(household refuses | contacted)
$= (0.62)(1 - 0.14) = 0.5332$

**b)** $P$(fail to contact household or contacting and not getting interview)

= $P$(fail to contact) + $P$(contact household)$P$(not getting interview | contacted)

= $(1 - 0.62) + (0.62)(1 - 0.14) = 0.9132$

**c)** The question in part b covers all possible occurrences *except* contacting the house and getting the interview.

$P$(failing to contact household or contacting and not getting the interview)

= $1 - P$(contacting the household and getting the interview)

= $1 - (0.62)(0.14) = 0.9132$

**47. M&M's.**

**a)** Since all of the events are disjoint (an M&M can't be two colors at once!), use the addition rule where applicable.

**1.** $P$(brown) = $1 - P$(not brown) = $1 - P$(yellow or red or orange or blue or green)
= $1 - (0.20 + 0.20 + 0.10 + 0.10 + 0.10) = 0.30$

**2.** $P$(yellow or orange) = $0.20 + 0.10 = 0.30$

**3.** $P$(not green) = $1 - P$(green) = $1 - 0.10 = 0.90$

**4.** $P$(striped) = 0

**b)** Since the events are independent (picking out one M&M doesn't affect the outcome of the next pick), the multiplication rule may be used.

**1.** $P$(all three are brown) = $(0.30)(0.30)(0.30) = 0.027$

**2.** $P$(the third one is the first one that is red) = $P$(not red and not red and red)
$\qquad\qquad\qquad\qquad\qquad\qquad\qquad = (0.80)(0.80)(0.20) = 0.128$

**3.** $P$(no yellow) = $P$(not yellow and not yellow and not yellow)
$\qquad\qquad\qquad = (0.80)(0.80)(0.80) = 0.512$

**4.** $P$(at least one is green) = $1 - P$(none are green) = $1 - (0.90)(0.90)(0.90) = 0.271$

**49. Disjoint or independent?**

**a)** For one draw, the events of getting a red M&M and getting an orange M&M are disjoint events. Your single draw cannot be both red and orange.

**b)** For two draws, the events of getting a red M&M on the first draw and a red M&M on the second draw are independent events. Knowing that the first draw is red does not influence the probability of getting a red M&M on the second draw.

    **c)** Disjoint events can never be independent. Once you know that one of a pair of disjoint events has occurred, the other one cannot occur, so its probability has become zero. For example, consider drawing one M&M. If it is red, it cannot possible be orange. Knowing that the M&M is red influences the probability that the M&M is orange. It's zero. The events are not independent.

## 51. Champion bowler.

Assuming each frame is independent of others, the multiplication rule may be used.

**a)** $P$(no strikes in 3 frames) = (0.30)(0.30)(0.30) = 0.027

**b)** $P$(makes first strike in the third frame) = (0.30)(0.30)(0.70) = 0.063

**c)** $P$(at least one strike in first three frames) = 1 – $P$(no strikes) = 1 – $(0.30)^3$ = 0.973

**d)** $P$(perfect game) = $(0.70)^{12} \approx 0.014$

## 53. Lights.

Assume that the defective light bulbs are distributed randomly to all stores so that the events can be considered independent. The multiplication rule may be used.

$P$(at least one of five bulbs is defective) = 1 – $P$(none are defective)

$$= 1 - (0.94)(0.94)(0.94)(0.94)(0.94) \approx 0.266$$

## 55. 9/11?

**a)** For any date with a valid three-digit date, the chance is 0.001, or 1 in 1000. For many dates in October through December, the probability is 0. For example, there is no way three digits will make 1015, to match October 15.

**b)** There are 65 days when the chance to match is 0. (October 10 through October 31, November 10 through November 30, and December 10 through December 31.) That leaves 300 days in a year (that is not a leap year) in which a match might occur.
$P$(no matches in 300 days) = $(0.999)^{300} \approx 0.741$.

**c)** $P$(at least one match in a year) = 1 – $P$(no matches in a year) = 1 – 0.741 ≈ 0.259

**d)** $P$(at least one match on 9/11 in one of the 50 states)
= 1 – $P$(no matches in 50 states) = 1 – $(0.999)^{50} \approx 0.049$

## 57. Global survey.

**a)** $P(USA) = \dfrac{1557}{7690} \approx 0.2025$

**b)** $P$(some high school or primary or less) $= \dfrac{4195}{7690} + \dfrac{1161}{7690} \approx 0.6965$

**c)**

$$P(\text{France or post-graduate}) = P(\text{France}) + P(\text{post-graduate}) - P(\text{both})$$

$$= \frac{1539}{7690} + \frac{379}{7690} - \frac{69}{7690} \approx 0.2404$$

**d)** $P(\text{France and primary school or less}) = \frac{309}{7690} \approx 0.0402$

## 59. Health.

Construct a two-way table of the conditional probabilities, including the marginal probabilities.

|  | Blood Pressure | | |
|---|---|---|---|
| Cholesterol | High | OK | Total |
| High | 0.11 | 0.21 | 0.32 |
| OK | 0.16 | 0.52 | 0.68 |
| Total | 0.27 | 0.73 | 1.00 |

**a)** $P(\text{both conditions}) = 0.11$

**b)** $P(\text{high BP}) = 0.11 + 0.16 = 0.27$

**c)** $P(\text{high chol.} \mid \text{high BP}) = \dfrac{P(\text{high chol. and high BP})}{P(\text{high BP})} = \dfrac{0.11}{0.27} \approx 0.407$

Consider only the High Blood Pressure column. Within this column, the probability of having high cholesterol is 0.11 out of a total of 0.27.

**d)** $P(\text{high BP} \mid \text{high chol.}) = \dfrac{P(\text{high BP and high chol.})}{P(\text{high chol.})} = \dfrac{0.11}{0.32} \approx 0.344$

This time, consider only the high cholesterol row. Within this row, the probability of having high blood pressure is 0.11, out of a total of 0.32.

## 61. Global survey, take 2.

**a)** $P(\text{USA and postgraduate work}) = \dfrac{84}{7690} \approx 0.011$

**b)** $P(\text{USA} \mid \text{post-graduate}) = \dfrac{84}{379} \approx 0.222$

**c)** $P(\text{post-graduate} \mid \text{USA}) = \dfrac{84}{1557} \approx 0.054$

**d)** $P(\text{primary} \mid \text{China}) = \dfrac{506}{1502} \approx 0.337$

**e)** $P(\text{China} \mid \text{primary}) = \dfrac{506}{1161} \approx 0.436$

### 63. Batteries.

Since batteries are not being replaced, use conditional probabilities throughout.

a)

$P(\text{the first two batteries are good}) = P(\text{good})P(\text{good})$

$$= \left(\frac{7}{12}\right)\left(\frac{6}{11}\right) \approx 0.318$$

b)

$P(\text{at least one of the first three batteries works})$

$= 1 - P(\text{none of the first three batteries work})$

$= 1 - \left[P(\text{no good})P(\text{no good})P(\text{no good})\right]$

$= 1 - \left(\frac{5}{12}\right)\left(\frac{4}{11}\right)\left(\frac{3}{10}\right) \approx 0.955$

c)

$P(\text{the first four batteries are good}) = P(\text{good})P(\text{good})P(\text{good})P(\text{good})$

$$= \left(\frac{7}{12}\right)\left(\frac{6}{11}\right)\left(\frac{5}{10}\right)\left(\frac{4}{9}\right) \approx 0.071$$

d)

$P(\text{pick five to find one good})$

$= P(\text{not good})P(\text{not good})P(\text{not good})P(\text{not good})P(\text{good})$

$= \left(\frac{5}{12}\right)\left(\frac{4}{11}\right)\left(\frac{3}{10}\right)\left(\frac{2}{9}\right)\left(\frac{7}{8}\right) \approx 0.009$

### 65. Eligibility.

a)

$P(\text{eligible}) = P(\text{stats}) + P(\text{computer}) - P(\text{both})$

$= 0.52 + 0.23 - 0.07$

$= 0.68$

68% of students are eligible for BioResearch, so
100 – 68 = 32% are ineligible.

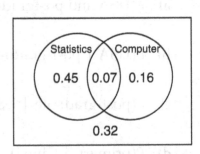

From the Venn, the region outside the circles represents those students who have
taken neither course, and are therefore ineligible for BioResearch.

**b)**

$$P(\text{computer course} \mid \text{statistics}) = \frac{P(\text{computer and statistics})}{P(\text{statistics})} = \frac{0.07}{0.52} \approx 0.135$$

From the Venn, consider only the region inside the Statistics circle. The probability of having taken a computer course is 0.07 out of a total of 0.52 (the entire Statistics circle).

**c)** Taking the two courses are not disjoint events, since they have outcomes in common. In fact, 7% of juniors have taken both courses.

**d)** Taking the two courses are not independent events. The overall probability that a junior has taken a computer course is 0.23. The probability that a junior has taken a computer course given that he or she has taken a statistics course is 0.135. If taking the two courses were independent events, these probabilities would be the same.

## 67. Unsafe food.

**a)** Using the Venn diagram, the probability that a tested chicken was not contaminated with either kind of bacteria is 33%.

**b)** Contamination with campylobacter and contamination with salmonella are not disjoint events, since 9% of chicken is contaminated with both.

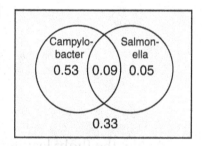

**c)** Contamination with campylobacter and contamination with salmonella may be independent events. The probability that a tested chicken is contaminated with campylobacter is 0.62. The probability that chicken contaminated with salmonella is also contaminated with campylobacter is 0.09/0.14 ≈ 0.64. If chicken is contaminated with salmonella, it is only slightly more likely to be contaminated with campylobacter than chicken in general. This difference could be attributed to expected variation due to sampling.

## 69. Men's health, again.

Consider the two-way table from a previous exercise.

High blood pressure and high cholesterol are not independent events. 28.8% of men with OK blood pressure have high cholesterol, while 40.7% of men with high blood pressure have high

|  | Blood Pressure | | |
|---|---|---|---|
| Cholesterol | | High | OK | Total |
| High | 0.11 | 0.21 | 0.32 |
| OK | 0.16 | 0.52 | 0.68 |
| Total | 0.27 | 0.73 | 1.00 |

cholesterol. If having high blood pressure and high cholesterol were independent, these percentages would be the same.

**71. Gender.**

According to the poll, party affiliation is not independent of sex.
Overall, $(32+41)/186 \approx 39.25\%$ of the respondents were Democrats. Of the men, only $32/94 \approx 34.04\%$ were Democrats.

**73. Luggage.**

Organize using a tree diagram.

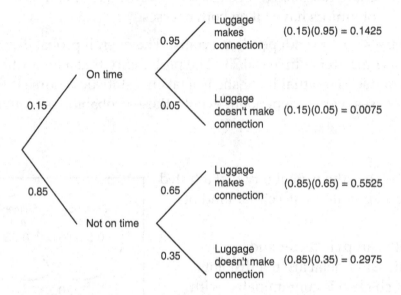

a) No, the flight leaving on time and the luggage making the connection are not independent events. The probability that the luggage makes the connection is dependent on whether or not the flight is on time. The probability is 0.95 if the flight is on time, and only 0.65 if it is not on time.

b) $P(\text{Luggage}) = P(\text{On time and Luggage}) + P(\text{Not on time and Luggage})$
$$= (0.15)(0.95) + (0.85)(0.65)$$
$$= 0.695$$

**75. Late luggage.**

Refer to the tree diagram constructed for Exercise 73.

$$P(\text{Not on time} \mid \text{No Lug.}) = \frac{P(\text{Not on time and No Luggage})}{P(\text{No Luggage})}$$

$$= \frac{(0.85)(0.35)}{(0.15)(0.05) + (0.85)(0.35)} \approx 0.975$$

If you pick Leah up at the Denver airport and her luggage is not there, the probability that her first flight was delayed is 0.975.

### 77. Absenteeism.

Organize the information in a tree diagram.

a) No, absenteeism is not independent of shift worked. The rate of absenteeism for the night shift is 2%, while the rate for the day shift is only 1%. If the two were independent, the percentages would be the same.

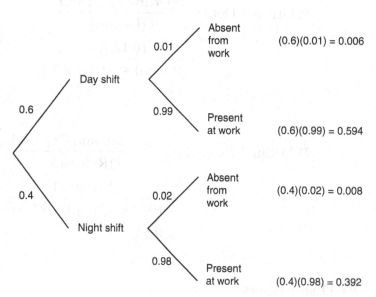

b)

$P(\text{Absent}) = P(\text{Day and Absent}) + P(\text{Night and Absent}) = (0.6)(0.01) + (0.4)(0.02) = 0.014$

The overall rate of absenteeism at this company is 1.4%.

### 79. Absenteeism, part II.

Refer to the tree diagram constructed for Exercise 77.

$$P(\text{Night} \mid \text{Absent}) = \frac{P(\text{Night and Absent})}{P(\text{Absent})} = \frac{(0.4)(0.02)}{(0.6)(0.01) + (0.4)(0.02)} \approx 0.571$$

Approximately 57.1% of the company's absenteeism occurs on the night shift.

### 81. Drunks.

Organize the information into a tree diagram.

a) $P(\text{Detain} \mid \text{Not Drinking}) = 0.2$

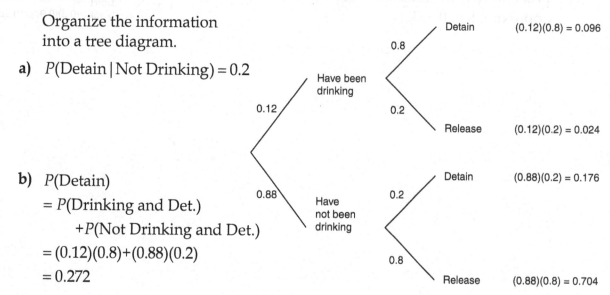

b) $P(\text{Detain})$
   $= P(\text{Drinking and Det.})$
   $\quad + P(\text{Not Drinking and Det.})$
   $= (0.12)(0.8) + (0.88)(0.2)$
   $= 0.272$

**c)**

$$P(\text{Drunk} \mid \text{Det.}) = \frac{P(\text{Drunk and Det.})}{P(\text{Detain})}$$

$$= \frac{(0.12)(0.8)}{(0.12)(0.8)+(0.88)(0.2)}$$

$$\approx 0.353$$

**d)**

$$P(\text{Drunk} \mid \text{Release}) = \frac{P(\text{Drunk and Release})}{P(\text{Release})}$$

$$= \frac{(0.12)(0.2)}{(0.12)(0.2)+(0.88)(0.8)}$$

$$\approx 0.033$$

**83. Dishwashers.**

Organize the information in a tree diagram.

$$P(\text{Chuck} \mid \text{Break})$$
$$= \frac{P(\text{Chuck and Break})}{P(\text{Break})}$$
$$= \frac{(0.3)(0.03)}{(0.4)(0.01)+(0.3)(0.01)+(0.3)(0.03)}$$
$$\approx 0.563$$

If you hear a dish break, the probability that Chuck is on the job is approximately 0.563.

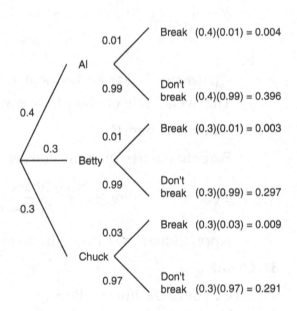

**85. HIV Testing.**

Organize the information in a tree diagram.

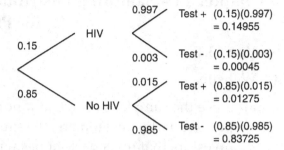

$$P(\text{No HIV} \mid \text{Test -}) = \frac{P(\text{No HIV and Test -})}{P(\text{Test -})}$$

$$= \frac{0.83725}{0.00045 + 0.83725}$$

$$\approx 0.999$$

The probability that a patient testing negative is truly free of HIV is about 99.9%.

## Chapter 13 – Sampling Distribution Models and Confidence Intervals for Proportions

**Section 13.1**

1. **Website.**

   a) Since the sample is drawn at random, and assuming that 200 investors is a small portion of their customers, the sampling distribution for the proportion of 200 investors that use smartphones will be unimodal and symmetric (roughly Normal).

   b) The center of the sampling distribution of the sample proportion is 0.36.

   c) The standard deviation of the sample proportion is $\sqrt{\dfrac{pq}{n}} = \sqrt{\dfrac{(0.36)(0.64)}{200}} \approx 0.034$.

3. **Send money.**

   All of the histograms are centered around $p = 0.05$. As $n$ gets larger, the shape of the histograms get more unimodal and symmetric, approaching a Normal model, while the variability in the sample proportions decreases.

5. **Living online.**

   a) This means that 56% of the 1060 teens in the sample said they go online several times per day. This is our best estimate of $p$, the proportion of all U.S. teens who would say they have done so.

   b) $SE(\hat{p}) = \sqrt{\dfrac{(0.56)(0.44)}{1060}} \approx 0.0152$

   c) Because we don't know $p$, we use $\hat{p}$ to estimate the standard deviation of the sampling distribution. The standard error is our estimate of the amount of variation in the sample proportion we expect to see from sample to sample when we ask 1060 teens whether they go online several times per day.

**Section 13.2**

7. **Marriage.**

   The data come from a random sample, so the randomization condition is met. We don't know the exact value of $p$, we can estimate it with $\hat{p}$.

   $n\hat{p} = (1500)(0.27) = 405$, and $n\hat{q} = (1500)(0.73) = 1095$. So, there are well over 10 successes and 10 failures, meeting the Success/Failure Condition. Since there are more than $(10)(1500) = 15,000$ adults in the United States, the 10% Condition is met. A Normal model is appropriate for the sampling distribution of the sample proportion.

## 9. Send more money.

**a)** The histogram for $n = 200$ looks quite unimodal and symmetric. We should be able to use the Normal model.

**b)** The Success/Failure condition requires $np$ and $nq$ to both be at least 10, which is not satisfied until $n = 200$ for $p = 0.05$. The theory supports the choice in part a.

## 11. Sample maximum.

**a)** A Normal model is not appropriate for the sampling distribution of the sample maximum. The histogram is skewed strongly to the left.

**b)** No. The 95% rule is based on the Normal model, and the Normal model is not appropriate here.

### Section 13.3

## 13. Still living online.

**a)** We are 95% confident that, if we were to ask all U.S. teens whether the go online several times per day, between 53% and 59% of them would say they do.

**b)** If we were to collect many random samples of 1060 teens, about 95% of the confidence intervals we construct would contain the proportion of all U.S. teens who say they go online several times per day.

### Section 13.4

## 15. Wrong direction.

**a)** $ME = z^* \times SE(\hat{p}) = z^* \times \sqrt{\dfrac{\hat{p}\hat{q}}{n}} = 1.645 \times \sqrt{\dfrac{(0.54)(0.46)}{2214}} \approx 0.017$ or 1.7%.

**b)** We are 90% confident that the observed proportion responding "Wrong Track" is within 0.017 of the population proportion.

### Section 13.5

## 17. Wrong direction again.

**a)** The sample is a simple random sample. Both $n\hat{p} = (2214)(0.54) = 1196$ and $n\hat{q} = (2214)(0.46) = 1018$ are at least ten. The sample size of 2214 is less than 10% of the population of all US voters. Conditions for the confidence interval are met.

**b)** The margin of error for 95% confidence would be larger. The critical value of 1.96 is greater than the 1.645 needed for 90% confidence.

**Section 13.6**

**19. Graduation.**

a)

$$ME = z^* \sqrt{\frac{\hat{p}\hat{q}}{n}}$$

$$0.06 = 1.645 \sqrt{\frac{(0.25)(0.75)}{n}}$$

$$n = \frac{(1.645)^2 (0.25)(0.75)}{(0.06)^2}$$

$$n \approx 141 \text{ people}$$

In order to estimate the proportion of non-graduates in the 25- to 30-year-old age group to within 6% with 90% confidence, we would need a sample of at least 141 people. All decimals in the final answer must be rounded up, to the next person.

(For a more cautious answer, let $\hat{p} = \hat{q} = 0.5$. This method results in a required sample of 188 people.)

b)

$$ME = z^* \sqrt{\frac{\hat{p}\hat{q}}{n}}$$

$$0.04 = 1.645 \sqrt{\frac{(0.25)(0.75)}{n}}$$

$$n = \frac{(1.645)^2 (0.25)(0.75)}{(0.04)^2}$$

$$n \approx 318 \text{ people}$$

In order to estimate the proportion of non-graduates in the 25- to 30-year-old age group to within 4% with 90% confidence, we would need a sample of at least 318 people. All decimals in the final answer must be rounded up, to the next person.

(For a more cautious answer, let $\hat{p} = \hat{q} = 0.5$. This method results in a required sample of 423 people.)

Alternatively, the margin of error is now 2/3 of the original, so the sample size must be increased by a factor of 9/4. $141(9/4) \approx 318$ people.

c)

$$ME = z^* \sqrt{\frac{\hat{p}\hat{q}}{n}}$$

$$0.03 = 1.645 \sqrt{\frac{(0.25)(0.75)}{n}}$$

$$n = \frac{(1.645)^2 (0.25)(0.75)}{(0.03)^2}$$

$$n \approx 564 \text{ people}$$

In order estimate the proportion of non-graduates in the 25- to 30-year-old age group to within 3% with 90% confidence, we would need a sample of at least 564 people. All decimals in the final answer must be rounded up, to the next person.

(For a more cautious answer, let $\hat{p} = \hat{q} = 0.5$. This method results in a required sample of 752 people.)

Alternatively, the margin of error is now half that of the original, so the sample size must be increased by a factor of 4. $141(4) \approx 564$ people.

**Chapter Exercises**

**21. Margin of error.**

The newscaster believes the true proportion of voters with a certain opinion is within 4% of the estimate, with some degree of confidence, perhaps 95% confidence.

**23. Conditions.**

a) *Population* – all cars; *sample* – 134 cars actually stopped at the checkpoint; $p$ – proportion of all cars with safety problems; $\hat{p}$ – proportion of cars in the sample that actually have safety problems (10.4%).
**Randomization condition:** This sample is not random, so hopefully the cars stopped are representative of cars in the area.
**10% condition:** The 134 cars stopped represent a small fraction of all cars, certainly less than 10%.
**Success/Failure condition:** $n\hat{p} = 14$ and $n\hat{q} = 120$ are both greater than 10, so the sample is large enough.
A one-proportion z-interval can be created for the proportion of all cars in the area with safety problems.

b) *Population* – the general public; *sample* – 602 viewers that logged on to the Web site;
$p$ –proportion of the general public that support prayer in school; $\hat{p}$ – proportion of viewers that logged on to the Web site and voted that support prayer in schools (81.1%).
**Randomization condition:** This sample is not random, but biased by voluntary response.
It would be very unwise to attempt to use this sample to infer anything about the opinion of the general public related to school prayer.

c) *Population* – parents at the school; *sample* – 380 parents who returned surveys; $p$ – proportion of all parents in favor of uniforms; $\hat{p}$ – proportion of those who responded that are in favor of uniforms (60%).
**Randomization condition:** This sample is not random, but rather biased by nonresponse. There may be lurking variables that affect the opinions of parents who return surveys (and the children who deliver them!).
It would be very unwise to attempt to use this sample to infer anything about the opinion of the parents about uniforms.

d) *Population* – all freshmen enrollees at the college (not just one year); *sample* – 1632 freshmen during the specified year; $p$ – proportion of all students who will graduate on time; $\hat{p}$ – proportion of on time graduate that year (85.05%).

**Randomization condition:** This sample is not random, but this year's freshmen class is probably representative of freshman classes in other years.

**10% condition:** The 1632 students in that years freshmen class represent less than 10% of all possible students.

**Success/Failure condition:** $n\hat{p} = 1388$ and $n\hat{q} = 244$ are both greater than 10, so the sample is large enough.

A one-proportion z-interval can be created for the proportion of freshmen that graduate on time from this college.

**10% condition:** The 309 employees represent less than 10% of all possible employees over many years.

**Success/Failure condition:** $n\hat{p} = 12$ and $n\hat{q} = 297$ are both greater than 10, so the sample is large enough.

A one-proportion z-interval can be created for the proportion of employees who are expected to suffer an injury on the job in future years, provided that this year is representative of future years.

## 25. Conclusions.

a) Not correct. This statement implies certainty. There is no level of confidence in the statement.

b) Not correct. Different samples will give different results. Many fewer than 95% of samples are expected to have *exactly* 88% on-time orders.

c) Not correct. A confidence interval should say something about the unknown population proportion, not the sample proportion in different samples.

d) Not correct. We *know* that 88% of the orders arrived on time. There is no need to make an interval for the sample proportion.

e) Not correct. The interval should be about the proportion of on-time orders, not the days.

## 27. Confidence intervals.

a) False. For a given sample size, higher confidence means a *larger* margin of error.

b) True. Larger samples lead to smaller standard errors, which lead to smaller margins of error.

c) True. Larger samples are less variable, which makes us more confident that a given confidence interval succeeds in catching the population proportion.

**d)** False. The margin of error decreases as the square root of the sample size increases. Halving the margin of error requires a sample four times as large as the original.

### 29. Cars.

We are 90% confident that between 29.9% and 47.0% of cars are made in Japan.

### 31. Mislabeled seafood.

**a)** $\hat{p} \pm z^* \sqrt{\dfrac{\hat{p}\hat{q}}{n}} = (0.33) \pm 1.960 \sqrt{\dfrac{(0.33)(0.67)}{1215}} = (0.304, 0.356)$

**b)** We are 95% confident that between 30.4% and 35.6% of all seafood packages purchased in the United States are mislabeled.

**c)** The size of the population is irrelevant. If *Consumer Reports* had a random sample, 95% of intervals generated by studies like this are expected to capture the true proportion of seafood packages that are mislabeled.

### 33. Baseball fans.

**a)** $ME = z^* \times SE(\hat{p}) = z^* \times \sqrt{\dfrac{\hat{p}\hat{q}}{n}} = 1.645 \times \sqrt{\dfrac{(0.48)(0.52)}{1006}} \approx 0.026$

**b)** We're 90% confident that this poll's estimate is within 2.6% of the true proportion of people who are baseball fans.

**c)** The margin of error for 99% confidence would be larger. To be more certain, we must be less precise.

**d)** $ME = z^* \times SE(\hat{p}) = z^* \times \sqrt{\dfrac{\hat{p}\hat{q}}{n}} = 2.576 \times \sqrt{\dfrac{(0.48)(0.52)}{1006}} \approx 0.041$

**e)** Smaller margins of error involve less confidence. The narrower the confidence interval, the less likely we are to believe that we have succeeded in capturing the true proportion of people who are baseball fans.

### 35. Contributions please.

**a)** **Randomization condition:** Letters were sent to a random sample of 100,000 potential donors.
**10% condition:** We assume that the potential donor list has more than 1,000,000 names.
**Success/Failure condition:** $n\hat{p} = 4781$ and $n\hat{q} = 95,219$ are both much greater than 10, so the sample is large enough.

$\hat{p} \pm z^* \sqrt{\dfrac{\hat{p}\hat{q}}{n}} = \left(\dfrac{4781}{100,000}\right) \pm 1.960 \sqrt{\dfrac{\left(\frac{4781}{100,000}\right)\left(\frac{95,219}{100,000}\right)}{100,000}} = (0.0465, 0.0491)$

We are 95% confident that the between 4.65% and 4.91% of potential donors would donate.

b) The confidence interval gives the set of plausible values with 95% confidence. Since 5% is above the interval, it seems to be a bit optimistic.

## 37. Teenage drivers.

a) **Randomization condition:** The insurance company randomly selected 582 accidents.
   **10% condition:** 582 accidents represent less than 10% of all accidents.
   **Success/Failure condition:** $n\hat{p} = 91$ and $n\hat{q} = 491$ are both greater than 10, so the sample is large enough.

   Since the conditions are met, we can use a one-proportion $z$-interval to estimate the percentage of accidents involving teenagers.

   $$\hat{p} \pm z^* \sqrt{\frac{\hat{p}\hat{q}}{n}} = \left(\frac{91}{582}\right) \pm 1.960 \sqrt{\frac{\left(\frac{91}{582}\right)\left(\frac{491}{582}\right)}{582}} = (12.7\%, 18.6\%)$$

b) We are 95% confident that between 12.7% and 18.6% of all accidents involve teenagers.

c) About 95% of random samples of size 582 will produce intervals that contain the true proportion of accidents involving teenagers.

d) Our confidence interval contradicts the assertion of the politician. The figure quoted by the politician, 1 out of every 5, or 20%, is above the interval.

## 39. Safe food.

The grocer can conclude nothing about the opinions of all his customers from this survey. Those customers who bothered to fill out the survey represent a voluntary response sample, consisting of people who felt strongly one way or another about irradiated food. The random condition was not met.

## 41. Death penalty, again.

a) There may be response bias based on the wording of the question.

b) $\hat{p} \pm z^* \sqrt{\frac{\hat{p}\hat{q}}{n}} = (0.585) \pm 1.960 \sqrt{\frac{(0.585)(0.415)}{1020}} = (56\%, 62\%)$

c) The margin of error based on the pooled sample is smaller, since the sample size is larger.

### 43. Rickets.

a) **Randomization condition:** The 2700 children were chosen at random.
**10% condition:** 2700 children are less than 10% of all English children.
**Success/Failure condition:** $n\hat{p} = (2700)(0.20) = 540$ and $n\hat{q} = (2700)(0.80) = 2160$ are both greater than 10, so the sample is large enough.

Since the conditions are met, we can use a one-proportion $z$-interval to estimate the proportion of the English children with vitamin D deficiency.

$$\hat{p} \pm z^* \sqrt{\frac{\hat{p}\hat{q}}{n}} = (0.20) \pm 2.326 \sqrt{\frac{(0.20)(0.80)}{2700}} = (18.2\%, 21.8\%)$$

b) We are 98% confident that between 18.2% and 21.8% of English children are deficient in vitamin D.

c) About 98% of random samples of size 2700 will produce confidence intervals that contain the true proportion of English children that are deficient in vitamin D.

### 45. Privacy or Security?

a) The confidence interval will be wider. The sample size is probably about one-sixth (17%) of the sample size of for all adults, so we would expect the confidence interval to be about two and a half times as wide.

b) The second poll's margin of error should be slightly wider. There are fewer "young" people (13%) in the sample than seniors (17%).

### 47. Deer ticks.

a) **Independence assumption:** Deer ticks are parasites. A deer carrying the parasite may spread it to others. Ticks may not be distributed evenly throughout the population.
**Randomization condition:** The sample is not random and may not represent all deer.
**10% condition:** 153 deer are less than 10% of all deer.
**Success/Failure condition:** $n\hat{p} = 32$ and $n\hat{q} = 121$ are both greater than 10, so the sample is large enough.

The conditions are not satisfied, so we should use caution when a one-proportion $z$-interval is used to estimate the proportion of deer carrying ticks.

$$\hat{p} \pm z^* \sqrt{\frac{\hat{p}\hat{q}}{n}} = \left(\frac{32}{153}\right) \pm 1.645 \sqrt{\frac{\left(\frac{32}{153}\right)\left(\frac{121}{153}\right)}{153}} = (15.5\%, 26.3\%)$$

We are 90% confident that between 15.5% and 26.3% of deer have ticks.

b) In order to cut the margin of error in half, they must sample 4 times as many deer.
$4(153) = 612$ deer.

**c)** The incidence of deer ticks is not plausibly independent, and the sample may not be representative of all deer, since females and young deer are usually not hunted.

### 49. Graduation, again.

$$ME = z^* \sqrt{\frac{\hat{p}\hat{q}}{n}}$$

$$0.02 = 1.960 \sqrt{\frac{(0.25)(0.75)}{n}}$$

$$n = \frac{(1.960)^2 (0.25)(0.75)}{(0.02)^2}$$

$$n \approx 1801 \text{ people}$$

In order to estimate the proportion of non-graduates in the 25-to 30-year-old age group to within 2% with 95% confidence, we would need a sample of at least 1801 people. All decimals in the final answer must be rounded up, to the next person.

(For a more cautious answer, let $\hat{p} = \hat{q} = 0.5$. This method results in a required sample of 2401 people.)

### 51. Pilot study.

$$ME = z^* \sqrt{\frac{\hat{p}\hat{q}}{n}}$$

$$0.03 = 1.645 \sqrt{\frac{(0.15)(0.85)}{n}}$$

$$n = \frac{(1.645)^2 (0.15)(0.85)}{(0.03)^2}$$

$$n \approx 384 \text{ cars}$$

Use $\hat{p} = \frac{9}{60} = 0.15$ from the pilot study as an estimate.

In order to estimate the percentage of cars with faulty emissions systems to within 3% with 90% confidence, the state's environmental agency will need a sample of at least 384 cars. All decimals in the final answer must be rounded up, to the next car.

### 53. Approval rating.

$$ME = z^* \sqrt{\frac{\hat{p}\hat{q}}{n}}$$

$$0.025 = z^* \sqrt{\frac{(0.65)(0.35)}{972}}$$

$$z^* = \frac{0.025}{\sqrt{\frac{(0.65)(0.35)}{972}}}$$

$$z^* \approx 1.634$$

Since $z^* \approx 1.634$, which is close to 1.645, the pollsters were probably using 90% confidence. The slight difference in the $z^*$ values is due to rounding of the governor's approval rating.

# Chapter 14 – Inferences About Means

**Section 14.1**

1. **Salmon.**

   **a)** The shipment of 4 salmon has $SD(\bar{y}) = \dfrac{\sigma}{\sqrt{n}} = \dfrac{2}{\sqrt{4}} = 1\,\text{pound.}$

   The shipment of 16 salmon has $SD(\bar{y}) = \dfrac{\sigma}{\sqrt{n}} = \dfrac{2}{\sqrt{16}} = 0.5\,\text{pounds.}$

   The shipment of 100 salmon has $SD(\bar{y}) = \dfrac{\sigma}{\sqrt{n}} = \dfrac{2}{\sqrt{100}} = 0.2\,\text{pounds.}$

   **b)** The Normal model would better characterize the shipping weight of the pallets than the shipping weight of the boxes, since the pallets contain a large number of salmon. The Central Limit Theorem tells us that the distribution of means (and therefore totals) approaches the Normal model, regardless of the underlying distribution. As samples get larger, the approximation gets better.

3. **Tips.**

   **a)** Since the distribution of tips is skewed to the right, we can't use the Normal model to determine the probability that a given party will tip at least $20.

   **b)** No. A sample of 4 parties is probably not a large enough sample for the CLT to allow us to use the Normal model to estimate the distribution of averages.

   **c)** A sample of 10 parties may not be large enough to allow the use of a Normal model to describe the distribution of averages. It would be risky to attempt to estimate the probability that his next 10 parties tip an average of $15. However, since the distribution of tips has $\mu = \$9.60$, with standard deviation $\sigma = \$5.40$, we still know that the mean of the sampling distribution model is $\mu_{\bar{y}} = \$9.60$ with standard deviation $SD(\bar{y}) = \dfrac{5.40}{\sqrt{10}} \approx \$1.71.$

   We don't know the exact shape of the distribution, but we can still assess the likelihood of specific means. A mean tip of $15 is over 3 standard deviations above the expected mean tip for 10 parties. That's not very likely to happen.

## 5. More tips.

**a) Randomization condition:** Assume that the tips from 40 parties can be considered a representative sample of all tips.

**Independence assumption:** It is reasonable to think that the tips are mutually independent, unless the service is particularly good or bad during this weekend.

**10% condition:** The tips of 40 parties certainly represent less than 10% of all tips.

**Large Enough Sample condition:** The sample of 40 parties is large enough.

The mean tip is $\mu = \$9.60$, with standard deviation $\sigma = \$5.40$. Since the conditions are satisfied, the CLT allows us to model the sampling distribution of $\bar{y}$ with a Normal model, with $\mu_{\bar{y}} = \$9.60$ and standard deviation

$$SD(\bar{y}) = \frac{5.40}{\sqrt{40}} \approx \$0.8538.$$

In order to earn at least $500, the waiter would have to average

$$\frac{500}{40} = \$12.50 \text{ per party.}$$

According to the Normal model, the probability that the waiter earns at least $500 in tips in a weekend is approximately 0.0003.

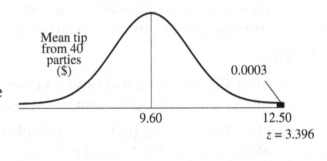

**b)** According to the Normal model, the waiter can expect to have a mean tip of about $10.6942, which corresponds to about $427.77 for 40 parties, in the best 10% of such weekends.

$$z = \frac{\bar{y} - \mu_{\bar{y}}}{SD(\bar{y})}$$

$$1.2816 = \frac{\bar{y} - 9.60}{\frac{5.40}{\sqrt{40}}}$$

$$\bar{y} \approx 10.6942$$

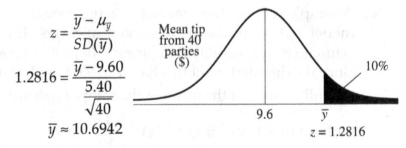

## Section 14.2

## 7. *t*-models, part I.

**a)** 1.74     **b)** 2.37

## 9. *t*-models, part III.

As the number of degrees of freedom increases, the shape and center of *t*-models do not change. The spread of *t*-models decreases as the number of degrees of freedom increases, and the shape of the distribution becomes closer to Normal.

## 11. Home sales.

a) The estimates of home value losses must be independent. This is verified using the Randomization condition, since the houses were randomly sampled. The distribution of home value losses must be Normal. A histogram of home value losses in the sample would be checked to verify this, using the Nearly Normal condition. Even if the histogram is not unimodal and symmetric, the sample size of 36 should allow for some small departures from Normality.

b) $\bar{y} \pm t^*_{n-1}\left(\dfrac{s}{\sqrt{n}}\right) = 9560 \pm t^*_{35}\left(\dfrac{1500}{\sqrt{36}}\right) \approx (9052.50, \ 10067.50)$

## Section 14.3

## 13. Home sales revisited.

We are 95% confident that the interval $9052.50 to $10,067.50 contains the true mean loss in home value. That is, 95% of all random samples of size 36 will contain the true mean.

## 15. Cattle.

a) Not correct. A confidence interval is for the mean weight gain of the population of all cows. It says nothing about individual cows. This interpretation also appears to imply that there is something special about the interval that was generated, when this interval is actually one of many that could have been generated, depending on the cows that were chosen for the sample.

b) Not correct. A confidence interval is for the mean weight gain of the population of all cows, not individual cows.

c) Not correct. We don't need a confidence interval about the average weight gain for cows in this study. We are certain that the mean weight gain of the cows in this study is 56 pounds. Confidence intervals are for the mean weight gain of the population of all cows.

d) Not correct. This statement implies that the average weight gain varies. It doesn't. We just don't know what it is, and we are trying to find it. The average weight gain is either between 45 and 67 pounds, or it isn't.

e) Not correct. This statement implies that there is something special about our interval, when this interval is actually one of many that could have been generated, depending on the cows that were chosen for the sample. The correct interpretation is that 95% of samples of this size will produce an interval that will contain the mean weight gain of the population of all cows.

### 17. Framingham revisited.

**a)** To find a 95% confidence interval for the mean, we need to exclude the most extreme 5% of bootstrap sample means, 2.5% from each side. Using the 2.5th percentile and the 97.5th percentile, the interval is 232.434 to 237.130.

**b)** We are 95% confident that the true mean cholesterol level is between 232.4 and 237.1 mg/dL.

**c)** We must assume that the individual values in the sample are independent. In this case, the assumption is valid, since the sample was random.

**Section 14.5**

### 19. Shoe sizes revisited.

The interval is a range of possible values for the mean shoe size. The average is not a value that any individual in the population will have, but an average of all the individuals.

### 21. Meal plan.

**a)** Not correct. The confidence interval is not about the individual students in the population.

**b)** Not correct. The confidence interval is not about individual students in the sample. In fact, we know exactly what these students spent, so there is no need to estimate.

**c)** Not correct. We know that the mean cost for students in this sample was $1467.

**d)** Not correct. A confidence interval is not about other sample means.

**e)** This is the correct interpretation of a confidence interval. It estimates a population parameter.

**Chapter Exercises**

### 23. Pulse rates.

**a)** We are 95% confident the interval 70.9 to 74.5 beats per minute contains the true mean heart rate.

**b)** The width of the interval is about 74.5 – 70.9 = 3.6 beats per minute. The margin of error is half of that, about 1.8 beats per minute.

**c)** The margin of error would have been larger. More confidence requires a larger critical value of $t$, which increases the margin of error.

### 25. CEO compensation.

We should be hesitant to trust this confidence interval, since the conditions for inference are not met. The distribution is highly skewed and there is an outlier.

**27. Cholesterol.**

a) We cannot apply the Central Limit Theorem to describe the distribution of cholesterol measurements. The Central Limit Theorem is about means and proportions, not individual observations.

b) **Randomization condition:** Although not specifically stated, it is safe to assume that the 2515 adults are a random sample of US adults.

**Nearly Normal Condition:** The distribution of the sample of cholesterol levels is skewed to the right, with several outliers on both ends. However, the sample size is very large, and the Central Limit Theorem will allow us to us a Normal model to describe the mean cholesterol level of samples of 2515 adults.

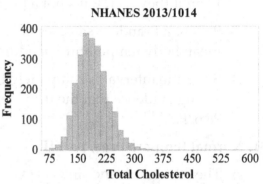

Therefore, the sampling distribution model for the mean cholesterol level of 2515 US adults is

$$N\left(188.9, \frac{41.6}{\sqrt{2515}}\right) \text{ or } N(188.9, 0.83).$$

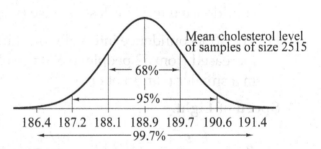

c) The Normal model is to the right.

**29. Normal temperature.**

a) **Randomization condition:** The adults were randomly selected.
**Nearly Normal condition:** The sample of 52 adults is large, and the histogram shows no serious skewness, outliers, or multiple modes.

The people in the sample had a mean temperature of 98.2846° and a standard deviation in temperature of 0.682379°. Since the conditions are satisfied, the sampling distribution of the mean can be modeled by a Student's $t$ model, with $52 - 1 = 51$ degrees of freedom. We will use a one-sample $t$-interval with 98% confidence for the mean body temperature.
(By hand, use $t^*_{50} \approx 2.403$ from the table.)

b) $\bar{y} \pm t^*_{n-1}\left(\dfrac{s}{\sqrt{n}}\right) = 98.2846 \pm t^*_{51}\left(\dfrac{0.682379}{\sqrt{52}}\right) \approx (98.06, 98.51)$

c) We are 98% confident that the interval 98.06°F to 98.51°F contains the true mean body temperature for adults. (If you calculated the interval by hand, using $t_{50}^* \approx 2.403$ from the table, your interval may be slightly different than intervals calculated using technology. With the rounding used here, they are identical. Even if they aren't, it's not a big deal.)

d) 98% of all random samples of size 52 will produce intervals that contain the true mean body temperature of adults.

e) Since the interval is completely below the body temperature of 98.6°F, there is strong evidence that the true mean body temperature of adults is lower than 98.6°F.

## 31. Normal temperatures, part II.

a) The 90% confidence interval would be narrower than the 98% confidence interval. We can be more precise with our interval when we are less confident.

b) The 98% confidence interval has a greater chance of containing the true mean body temperature of adults than the 90% confidence interval, but the 98% confidence interval is less precise (wider) than the 90% confidence interval.

c) The 98% confidence interval would be narrower if the sample size were increased from 52 people to 500 people. The smaller standard error would result in a smaller margin of error.

## 33. Speed of Light.

a) $\bar{y} \pm t_{n-1}^* \left( \dfrac{s}{\sqrt{n}} \right) = 756.22 \pm t_{22}^* \left( \dfrac{107.12}{\sqrt{23}} \right) \approx (709.9, \ 802.5)$

b) We are 95% confident that the interval 299,709.9 to 299,802.5 km/sec contains the speed of light.

c) We have assumed that the measurements are independent of each other and that the distribution of the population of all possible measurements is Normal. The assumption of independence seems reasonable, but it might be a good idea to look at a display of the measurements made by Michelson to verify that the Nearly Normal Condition is satisfied.

## 35. Flight on time 2016.

a) **Randomization condition:** Since there is no time trend, the monthly on-time departure rates should be independent. This is not a random sample, but should be representative.
**Nearly Normal condition:** The histogram looks unimodal, and slightly skewed to the left. Since the sample size is 270, this should not be of concern.

**b)** The on-time departure rates in the sample had a mean of 78.099%, and a standard deviation in of 5.010%. Since the conditions have been satisfied, construct a one-sample $t$-interval, with $270 - 1 = 269$ degrees of freedom, at 90% confidence.

$$\bar{y} \pm t_{n-1}^{*}\left(\frac{s}{\sqrt{n}}\right) = 78.099 \pm t_{269}^{*}\left(\frac{5.010}{\sqrt{270}}\right) \approx (77.596, 78.602)$$

**c)** We are 90% confident that the interval from 77.60% to 78.60% contains the true mean monthly percentage of on-time flight departures.

**37. Farmed salmon, second look.**

The 95% confidence interval lies entirely above the 0.08 ppm limit. This is evidence that mirex contamination is too high and consistent with rejecting the null hypothesis. We used an upper-tail test, so the $P$-value should be smaller than $\frac{1}{2}(1 - 0.95) = 0.025$, and it was.

**39. Pizza.**

Because even the lower bound of the confidence interval is above 220 mg/dL, we are quite confident that the mean cholesterol level for those who eat frozen pizza is a level that indicates a health risk.

**41. Fuel economy 2016 revisited.**

**a) Randomization condition:** The 35 cars were not selected randomly. We will have to assume that they are representative of all 2016 automobiles. **Nearly Normal condition:** The distribution doesn't appear to be unimodal and symmetric, but the sample size is reasonably large.

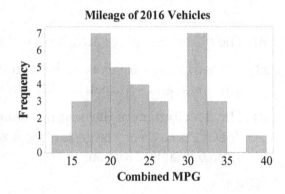

The mileages in the sample had a mean of 24.3429 mpg, and a standard deviation in of 6.53021 mpg. Since the conditions have been satisfied, construct a one-sample $t$-interval, with $35 - 1 = 34$ degrees of freedom, at 95% confidence.

$$\bar{y} \pm t_{n-1}^{*}\left(\frac{s}{\sqrt{n}}\right) = 24.3429 \pm t_{34}^{*}\left(\frac{6.53021}{\sqrt{35}}\right) \approx (22.1, 26.6)$$

We are 95% confident that the interval from 22.1 to 26.6 contains the true mean mileage of 2016 automobiles.

b) The data is a mix of small, mid-size, and large vehicles. Our histogram provides some evidence that there may be at least two distinct groups with regards to mileage. Without knowing how the data were selected, we are cautious about generalizing to all 2016 cars.

**43. Waist size.**

a) The distribution of waist size of 250 men is unimodal and slightly skewed to the right. A typical waist size is approximately 36 inches, and the standard deviation in waist sizes is approximately 4 inches.

b) All of the histograms show distributions of sample means centered near 36 inches. As $n$ gets larger the histograms approach the Normal model in shape, and the variability in the sample means decreases. The histograms are fairly Normal by the time the sample reaches size 5.

**45. Waist size revisited.**

a)

| $n$ | Observed mean | Theoretical mean | Observed st. dev. | Theoretical standard deviation |
|---|---|---|---|---|
| 2 | 36.314 | 36.33 | 2.855 | $4.019 / \sqrt{2} \approx 2.842$ |
| 5 | 36.314 | 36.33 | 1.805 | $4.019 / \sqrt{5} \approx 1.797$ |
| 10 | 36.341 | 36.33 | 1.276 | $4.019 / \sqrt{10} \approx 1.271$ |
| 20 | 36.339 | 36.33 | 0.895 | $4.019 / \sqrt{20} \approx 0.899$ |

b) The observed values are all very close to the theoretical values.

c) For samples as small as 5, the sampling distribution of sample means is unimodal and symmetric. The Normal model would be appropriate.

d) The distribution of the original data is nearly unimodal and symmetric, so it doesn't take a very large sample size for the distribution of sample means to be approximately Normal.

**47. GPAs.**

**Randomization condition:** Assume that the students are randomly assigned to seminars.

**Independence assumption:** It is reasonable to think that GPAs for randomly selected students are mutually independent.

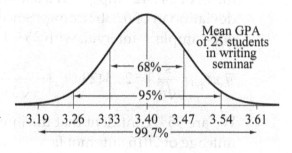

**Nearly Normal condition:** The distribution of GPAs is roughly unimodal and symmetric, so the sample of 25 students is large enough.

The mean GPA for the freshmen was $\mu = 3.4$, with standard deviation $\sigma = 0.35$. Since the conditions are met, the Central Limit Theorem tells us that we can model the sampling distribution of the mean GPA with a Normal model, with $\mu_{\bar{y}} = 3.4$ and standard deviation $SD(\bar{y}) = \dfrac{0.35}{\sqrt{25}} \approx 0.07$.

The sampling distribution model for the sample mean GPA is approximately $N(3.4, 0.07)$.

**49. Lucky spot?**

**a)** Smaller outlets have more variability than the larger outlets, just as the Central Limit Theorem predicts.

**b)** If the lottery is truly random, all outlets are equally likely to sell winning tickets.

**51. Pregnancy.**

**a)**

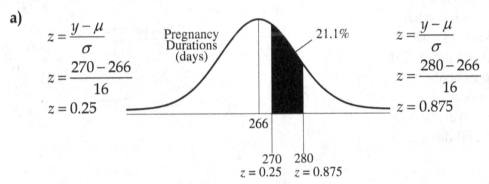

$$z = \frac{y - \mu}{\sigma}$$
$$z = \frac{270 - 266}{16}$$
$$z = 0.25$$

$$z = \frac{y - \mu}{\sigma}$$
$$z = \frac{280 - 266}{16}$$
$$z = 0.875$$

According to the Normal model, approximately 21.1% of all pregnancies are expected to last between 270 and 280 days.

**b)**

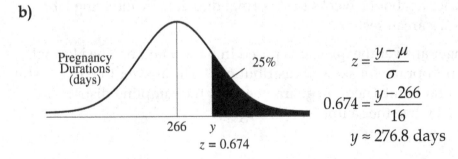

$$z = \frac{y - \mu}{\sigma}$$
$$0.674 = \frac{y - 266}{16}$$
$$y \approx 276.8 \text{ days}$$

According to the Normal model, the longest 25% of pregnancies are expected to last approximately 276.8 days or more.

c) **Randomization condition:** Assume that the 60 women the doctor is treating can be considered a representative sample of all pregnant women.

**Independence assumption:** It is reasonable to think that the durations of the patients' pregnancies are mutually independent.

**Nearly Normal condition:** The distribution of pregnancy durations is Normal.

The mean duration of the pregnancies was $\mu = 266$ days, with standard deviation $\sigma = 16$ days. Since the distribution of pregnancy durations is Normal, we can model the sampling distribution of the mean pregnancy duration with a Normal model, with $\mu_{\bar{y}} = 266$ days and standard deviation

$$SD(\bar{y}) = \frac{16}{\sqrt{60}} \approx 2.07 \text{ days}.$$

d) According to the Normal model, with mean 266 days and standard deviation 2.07 days, the probability that the mean pregnancy duration is less than 260 days is 0.002.

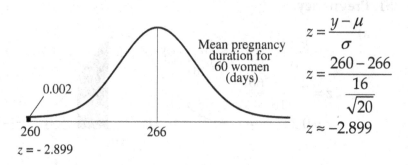

$$z = \frac{y - \mu}{\sigma}$$

$$z = \frac{260 - 266}{\frac{16}{\sqrt{20}}}$$

$$z \approx -2.899$$

## 53. Pregnant again.

a) The distribution of pregnancy durations may be skewed to the left since there are more premature births than very long pregnancies. Modern practice of medicine stops pregnancies at about 2 weeks past normal due date by inducing labor or performing a Caesarean section.

b) We can no longer answer the questions posed in parts a and b. The Normal model is not appropriate for skewed distributions. The answer to part c is still valid. The Central Limit Theorem guarantees that the sampling distribution model is Normal when the sample size is large.

## 55. Ruffles.

a) **Randomization condition:** The 6 bags were not selected at random, but it is reasonable to think that these bags are representative of all bags of chips.
**Nearly Normal condition:** The histogram of the weights of chips in the sample is nearly normal.

b) $\bar{y} \approx 28.78$ grams, $s \approx 0.40$ grams

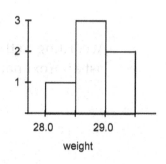

c) Since the conditions for inference have been satisfied, use a one-sample *t*-interval, with $6 - 1 = 5$ degrees of freedom, at 95% confidence.

$$\bar{y} \pm t^*_{n-1}\left(\frac{s}{\sqrt{n}}\right) = 28.78 \pm t^*_5\left(\frac{0.40}{\sqrt{6}}\right) \approx (28.36, 29.21)$$

d) We are 95% confident that the mean weight of the contents of Ruffles bags is between 28.36 and 29.21 grams.

e) Since the interval is above the stated weight of 28.3 grams, there is evidence that the company is filling the bags to more than the stated weight, on average.

f) The sample size of 6 bags is too small to provide a good basis for a bootstrap confidence interval.

**57. Popcorn.**

**Randomization condition:** The 8 bags were randomly selected.
**Nearly Normal condition:** The histogram of the percentage of unpopped kernels is unimodal and roughly symmetric.

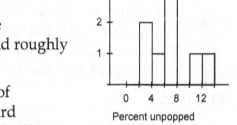

Percent unpopped

The bags in the sample had a mean percentage of unpopped kernels of 6.775 percent and a standard deviation in percentage of unpopped kernels of 3.637 percent. Since the conditions for inference are satisfied, we can use a one-sample *t*-interval, with $8 - 1 = 7$ degrees of freedom, at 95% confidence.

$$\bar{y} \pm t^*_{n-1}\left(\frac{s}{\sqrt{n}}\right) = 6.775 \pm t^*_7\left(\frac{3.637}{\sqrt{8}}\right) \approx (3.7344, 9.8156)$$

We are 95% confident that the true mean percentage of unpopped kernels is contained in the interval 3.73% to 9.82%. Since 10% is not contained in the interval, there is evidence that Yvon Hopps has met his goal of an average of no more than 10% unpopped kernels.

**59. Chips ahoy.**

a) **Randomization condition:** The bags of cookies were randomly selected.
**Nearly Normal condition:** The Normal probability plot is reasonably straight, and the histogram of the number of chips per bag is unimodal and symmetric.

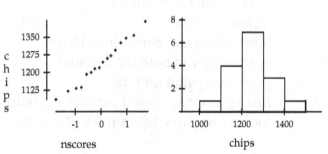

**b)** The bags in the sample had a mean number of chips of 1238.19, and a standard deviation of 94.282 chips. Since the conditions for inference have been satisfied, use a one-sample *t*-interval, with 16 − 1 = 15 degrees of freedom, at 95% confidence.

$$\bar{y} \pm t^*_{n-1}\left(\frac{s}{\sqrt{n}}\right) = 1238.19 \pm t^*_{15}\left(\frac{94.282}{\sqrt{16}}\right) \approx (1187.9,\ 1288.4)$$

We are 95% confident that the mean number of chips in an 18-ounce bag of Chips Ahoy cookies is between 1187.9 and 1288.4.

**c)** Since the confidence interval is well above 1000, there is strong evidence that the mean number of chips per bag is well above 1000.

However, since the "1000 Chip Challenge" is about individual bags, not means, the claim made by Nabisco may not be true. If the mean was around 1188 chips, the low end of our confidence interval, and the standard deviation of the population was about 94 chips, our best estimate obtained from our sample, a bag containing 1000 chips would be about 2 standard deviations below the mean. This is not likely to happen, but not an outrageous occurrence. These data do not provide evidence that the "1000 Chip Challenge" is true.

**61. Maze.**

**a)** The rats in the sample finished the maze with a mean time of 52.21 seconds and a standard deviation in times of 13.5646 seconds. Since the conditions for inference are satisfied, we can construct a one-sample *t*-interval, with 21 − 1 = 20 degrees of freedom, at 95% confidence.

$$\bar{y} \pm t^*_{n-1}\left(\frac{s}{\sqrt{n}}\right) = 52.21 \pm t^*_{20}\left(\frac{13.5646}{\sqrt{21}}\right) \approx (46.03,\ 58.38)$$

We are 95% confident that the true mean maze completion time is contained in the interval 46.03 seconds to 58.38 seconds.

**b)** **Independence assumption:** It is reasonable to think that the rats' times will be independent, as long as the times are for different rats.

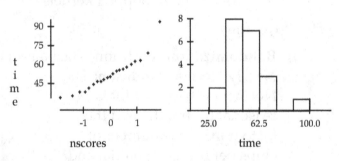

**Nearly Normal condition:** There is an outlier in both the Normal probability plot and the histogram that should probably be eliminated before continuing the test. One rat took a long time to complete the maze.

c) Without the outlier, the rats in the sample finished the maze with a mean time of 50.13 seconds and standard deviation in times of 9.90 seconds. Since the conditions for inference are satisfied, we can construct a one-sample $t$-interval, with $20 - 1 = 19$ degrees of freedom, at 95% confidence.

$$\bar{y} \pm t^*_{n-1}\left(\frac{s}{\sqrt{n}}\right) = 50.13 \pm t^*_{19}\left(\frac{9.90}{\sqrt{20}}\right) \approx (45.49, 54.77)$$

We are 95% confident that the true mean maze completion time is contained in the interval 45.49 seconds to 54.77 seconds.

d) According to both tests, there is evidence that the mean time required for rats to complete the maze is different than 60 seconds. The maze does not meet the "one-minute average" requirement. It should be noted that the test without the outlier is the appropriate test. The one slow rat made the mean time required seem much higher than it probably was.

## 63 Golf Drives III 2015.

a) $\bar{y} \pm t^*_{n-1}\left(\frac{s}{\sqrt{n}}\right) = 288.69 \pm t^*_{198}\left(\frac{9.28}{\sqrt{199}}\right) \approx (287.39, 289.99)$

b) These data are not a random sample of golfers. The top professionals are not representative of all golfers and were not selected at random. We might consider the 2016 data to represent the population of all professional golfers, past, present, and future.

c) The data are means for each golfer, so they are less variable than if we looked at separate drives, and inference is invalid.

## Chapter 15 – Testing Hypotheses

### Section 15.1

**1. Better than aspirin?**

The new drug is not more effective than aspirin, and reduces the risk of heart attack by 44%. ($p = 0.44$)

**3. Parameters and hypotheses.**

**a)** Let $p =$ probability of winning on a slot machine.
$H_0 : p = 0.01$ vs. $H_A : p \neq 0.01$

**b)** Let $\mu =$ mean spending per customer this year.
$H_0 : \mu = \$35.32$ vs. $H_A : \mu \neq \$35.32$

**c)** Let $p =$ proportion of patients cured by the new drug.
$H_0 : p = 0.3$ vs. $H_A : p \neq 0.3$

**d)** Let $p =$ proportion of clients now using the website.
$H_0 : p = 0.4$ vs. $H_A : p \neq 0.4$

### Section 15.2

**5. Better than aspirin again?**

**a)** The alternative to the null hypothesis is one-sided. They are interested in discovering only if their drug is more effective than aspirin, not if it is less effective than aspirin.

**b)** Since the *P*-value of 0.0028 is low, reject the null hypothesis. There is evidence that the new drug is more effective than aspirin.

**c)** Since the *P*-value of 0.28 is high, fail to reject the null hypothesis. There is not sufficient evidence to conclude that the new drug is more effective than aspirin.

### Section 15.3

**7. Hispanic origin.**

**a)** $H_0 :$ The proportion of people in the county that are of Hispanic or Latino origin is 0.16. ($p = 0.16$)

$H_A :$ The proportion of people in the county that are of Hispanic or Latino origin is different from 0.16. ($p \neq 0.16$)

b) **Randomization condition:** The 437 county residents were a random sample of all county residents.
**10% condition:** 437 is likely to be less than 10% of all county residents.

**Success/Failure condition:** $np_0 = (437)(0.16) = 69.92$ and $nq_0 = (437)(0.84) = 367.08$ are both greater than 10, so the sample is large enough.

Since the conditions are met, we will model the sampling distribution of $\hat{p}$ with a Normal model and perform a one-proportion z-test.

c) $\hat{p} = \dfrac{44}{437} = 0.101$

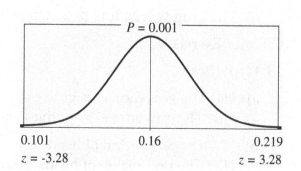

$$SD(\hat{p}) = \sqrt{\dfrac{p_0 q_0}{n}} = \sqrt{\dfrac{(0.16)(0.84)}{437}} \approx 0.018$$

$$z = \dfrac{\hat{p} - p_0}{SD(\hat{p})} = \dfrac{0.101 - 0.16}{0.018} = -3.28$$

$$\text{P-value} = 2 \cdot P(z < -3.28) = 0.001$$

d) Since the P-value = 0.001 is so low, reject the null hypothesis. There is evidence that the Hispanic/Latino population in this county differs from that as the nation as a whole. These data suggest that the proportion of Hispanic/Latino residents is, in fact, lower than the national proportion.

## Section 15.4

9. **GRE performance again.**

   a) An increase in the mean score would mean that the mean difference (After − Before) is positive.

      $H_0$ : The mean of the differences in GRE scores (After − Before) is zero. $\left(\mu_{diff} = 0\right)$

      $H_A$ : The mean of the differences in GRE scores (After − Before) is greater than zero. $\left(\mu_{diff} > 0\right)$

   b) Since the P-value, 0.65, is high, there is not convincing evidence that the mean difference in GRE scores (After − Before) is greater than zero. There is no evidence that the course is effective.

   c) Because the P-value is greater than 0.5 and the alternative is one-sided (>0), we can conclude that the actual mean difference was less than 0. (If it had been positive, the probability to the right of that value would have to be less than 0.5.) This means that, in the sample of customers, the scores generally decreased after the course.

**11. Pizza.**

If in fact the mean cholesterol of pizza eaters does not indicate a health risk, then only 7 out of every 100 samples would be expected to have mean cholesterol as high or higher than the mean cholesterol observed in the sample.

**Section 15.5**

**13. Bad medicine.**

a) The drug may not be approved for use, and people would miss out on a beneficial product.

b) The drug will go into production and people will suffer the side effect.

**Chapter Exercises**

**15. Hypotheses.**

a) $H_0$ : The governor's "negatives" are 30%. ($p = 0.30$)
$H_A$ : The governor's "negatives" are less than 30%. ($p < 0.30$)

b) $H_0$ : The proportion of heads is 50%. ($p = 0.50$)
$H_A$ : The proportion of heads is not 50%. ($p \neq 0.50$)

c) $H_0$ : The proportion of people who quit smoking is 20%. ($p = 0.20$)
$H_A$ : The proportion of people who quit smoking is greater than 20%. ($p > 0.20$)

**17. Negatives.**

Statement d is the correct interpretation of a $P$-value. It talks about the probability of seeing the data, not the probability of the hypotheses.

**19. Relief.**

It is *not* reasonable to conclude that the new formula and the old one are equally effective. Furthermore, our inability to make that conclusion has nothing to do with the $P$-value. We can not prove the null hypothesis (that the new formula and the old formula are equally effective), but can only fail to find evidence that would cause us to reject it. All we can say about this $P$-value is that there is a 27% chance of seeing the observed effectiveness from natural sampling variation if the new formula and the old one are equally effective.

**21. He cheats?**

a) Two losses in a row aren't convincing. There is a 25% chance of losing twice in a row, and that is not unusual.

b) If the process is fair, three losses in a row can be expected to happen about 12.5% of the time. $(0.5)(0.5)(0.5) = 0.125$.

c) Three losses in a row is still not a convincing occurrence. We'd expect that to happen about once every eight times we tossed a coin three times.

**d)** Answers may vary. Maybe 5 times would be convincing. The chances of 5 losses in a row are only 1 in 32, which seems unusual.

## 23. Smartphones.

**1)** Null and alternative hypotheses should involve $p$, not $\hat{p}$.

**2)** The question is about *failing* to meet the goal. $H_A$ should be $p < 0.96$.

**3)** The student failed to check $nq_0 = (200)(0.04) = 8$. Since $nq_0 < 10$, the Success/Failure condition is violated. Similarly, the 10% Condition is not verified.

**4)** $SD(\hat{p}) = \sqrt{\dfrac{pq}{n}} = \sqrt{\dfrac{(0.96)(0.04)}{200}} \approx 0.014$. The student used $\hat{p}$ and $\hat{q}$.

**5)** Value of $z$ is incorrect. The correct value is $z = \dfrac{0.94 - 0.96}{0.014} \approx -1.43$.

**6)** *P*-value is incorrect. $P = P(z < -1.43) = 0.076$

**7)** For the *P*-value given, an incorrect conclusion is drawn. A *P*-value of 0.12 provides no evidence that the new system has failed to meet the goal. The correct conclusion for the corrected *P*-value is: Since the *P*-value of 0.076 is fairly low, there is weak evidence that the new system has failed to meet the goal.

## 25. Dowsing.

**a)** $H_0$ : The percentage of successful wells drilled by the dowser is 30%. ($p = 0.30$)
$H_A$ : The percentage of successful wells drilled is greater than 30%. ($p > 0.30$)

**b)** **Independence assumption:** There is no reason to think that finding water in one well will affect the probability that water is found in another, unless the wells are close enough to be fed by the same underground water source.
**Randomization condition:** This sample is not random, so hopefully the customers you check with are representative of all of the dowser's customers.
**10% condition:** The 80 customers sampled may be considered less than 10% of all possible customers.
**Success/Failure condition:** $np_0 = (80)(0.30) = 24$ and $nq_0 = (80)(0.70) = 56$ are both greater than 10, so the sample is large enough.

**c)** The sample of customers may not be representative of all customers, so we will proceed cautiously. A Normal model can be used to model the sampling distribution of the proportion, with $\mu_{\hat{p}} = p_0 = 0.30$ and

$$SD(\hat{p}) = \sqrt{\dfrac{p_0 q_0}{n}} = \sqrt{\dfrac{(0.30)(0.70)}{80}} \approx 0.0512 \,.$$

We can perform a one-proportion z-test. The observed proportion of successful wells is $\hat{p} = \dfrac{27}{80} = 0.3375$.

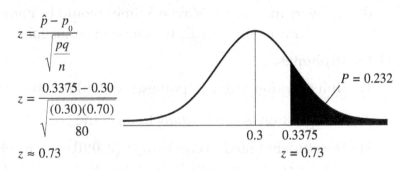

$$z = \frac{\hat{p} - p_0}{\sqrt{\dfrac{pq}{n}}}$$

$$z = \frac{0.3375 - 0.30}{\sqrt{\dfrac{(0.30)(0.70)}{80}}}$$

$$z \approx 0.73$$

$P = 0.232$

0.3    0.3375
$z = 0.73$

**d)** If his dowsing has the same success rate as standard drilling methods, there is more than a 23% chance of seeing results as good as those of the dowser, or better, by natural sampling variation.

**e)** With a high *P*-value of 0.232, we fail to reject the null hypothesis. There is no evidence to suggest that the dowser has a success rate any higher than 30%.

**27. Absentees.**

**a)** $H_0$ : The percentage of students in 2000 with perfect attendance the previous month is 34% ($p = 0.34$)
$H_A$ : The percentage of students in 2000 with perfect attendance the previous month is different from 34% ($p \neq 0.34$)

**b)** **Randomization condition:** Although not specifically stated, we can assume that the National Center for Educational Statistics used random sampling.
**10% condition:** The 8302 students are less than 10% of all students.
**Success/Failure condition:** $np_0 = (8302)(0.34) = 2822.68$ and $nq_0 = (8302)(0.66) = 5479.32$ are both greater than 10, so the sample is large enough.

**c)** Since the conditions for inference are met, a Normal model can be used to model the sampling distribution of the proportion, with $\mu_{\hat{p}} = p_0 = 0.34$ and

$$SD(\hat{p}) = \sqrt{\frac{p_0 q_0}{n}} = \sqrt{\frac{(0.34)(0.66)}{8302}} \approx 0.0052$$

We can perform a two-tailed one-proportion z-test. The observed proportion of perfect attendees is $\hat{p} = 0.33$.

**d)** With a *P*-value of 0.0544, we reject the null hypothesis. There is some evidence to suggest that the percentage of students with perfect attendance in the previous month has changed in 2000.

$$z = \frac{\hat{p} - p_0}{\sqrt{\dfrac{pq}{n}}}$$

$$z = \frac{0.33 - 0.34}{\sqrt{\dfrac{(0.34)(0.66)}{8302}}}$$

$$z \approx -1.923$$

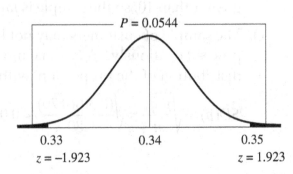

$P = 0.0544$

0.33    0.34    0.35
$z = -1.923$    $z = 1.923$

e) This result is not meaningful.  A difference this small, although statistically significant, is of little practical significance.

## 29. Contributions, please II.

a) $H_0$ : The contribution rate is 5% ($p = 0.05$)
   $H_A$ : The contribution rate is less than 5% ($p < 0.05$)

b) **Randomization condition:** Potential donors were randomly selected.
   **10% condition:** We will assume the entire mailing list has over 1,000,000 names.
   **Success/Failure condition:** $np_0 = 5000$ and $nq_0 = 95,000$ are both greater than 10, so the sample is large enough.

   The conditions have been satisfied, so a Normal model can be used to model the sampling distribution of the proportion, with $\mu_{\hat{p}} = p_0 = 0.05$ and

   $$SD(\hat{p}) = \sqrt{\frac{p_0 q_0}{n}} = \sqrt{\frac{(0.05)(0.95)}{100,000}} \approx 0.0007 \,.$$

   We can perform a one-proportion $z$-test.  The observed contribution rate is
   $$\hat{p} = \frac{4781}{100,000} = 0.04781 \,.$$

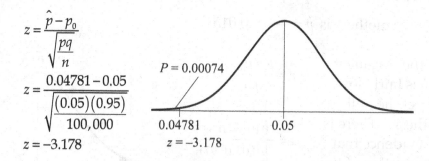

$$z = \frac{\hat{p} - p_0}{\sqrt{\dfrac{pq}{n}}}$$

$$z = \frac{0.04781 - 0.05}{\sqrt{\dfrac{(0.05)(0.95)}{100,000}}}$$

$$z = -3.178$$

$P = 0.00074$

$0.04781$     $0.05$
$z = -3.178$

c) Since the $P$-value $= 0.00074$ is low, we reject the null hypothesis.  There is strong evidence that contribution rate for all potential donors is lower than 5%.

## 31. Pollution.

$H_0$ : The percentage of cars with faulty emissions is 20%.  ($p = 0.20$)
$H_A$ : The percentage of cars with faulty emissions is  greater than 20%. ($p > 0.20$)

Two conditions are not satisfied.  22 is greater than 10% of the population of 150 cars, and $np_0 = (22)(0.20) = 4.4$, which is not greater than 10.  It's not advisable to proceed with a test.

### 33. Twins.

$H_0$ : The percentage of twin births to teenage girls is 3%. ($p = 0.03$)
$H_A$ : The percentage of twin births to teenage girls differs from 3%. ($p \neq 0.03$)

**Independence assumption:** One mother having twins will not affect another. Observations are plausibly independent.
**Randomization condition:** This sample may not be random, but it is reasonable to think that this hospital has a representative sample of teen mothers, with regards to twin births.
**10% condition:** The sample of 469 teenage mothers is less than 10% of all such mothers.
**Success/Failure condition:** $np_0 = (469)(0.03) = 14.07$ and $nq_0 = (469)(0.97) = 454.93$ are both greater than 10, so the sample is large enough.

The conditions have been satisfied, so a Normal model can be used to model the sampling distribution of the proportion, with $\mu_{\hat{p}} = p_0 = 0.03$ and

$$SD(\hat{p}) = \sqrt{\frac{p_0 q_0}{n}} = \sqrt{\frac{(0.03)(0.97)}{469}} \approx 0.0079 .$$

We can perform a one-proportion $z$-test. The observed proportion of twin births to teenage mothers is $\hat{p} = \dfrac{7}{469} \approx 0.015 .$

Since the $P$-value = 0.0556 is fairly low, we reject the null hypothesis. There is some evidence that the proportion of twin births for teenage mothers at this large city hospital is lower than the proportion of twin births for all mothers.

$$z = \frac{\hat{p} - p_0}{\sqrt{\dfrac{pq}{n}}}$$

$$z = \frac{0.015 - 0.03}{\sqrt{\dfrac{(0.03)(0.97)}{469}}}$$

$$z \approx -1.91$$

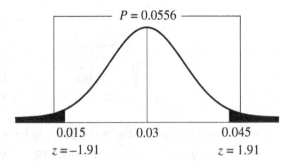

### 35. WebZine.

$H_0$ : The percentage of readers interested in an online edition is 25%. ($p = 0.25$)
$H_A$ : The percentage of readers interested is greater than 25%. ($p > 0.25$)

**Randomization condition:** The magazine conducted an SRS of 500 current readers.
**10% condition:** 500 readers are less than 10% of all potential subscribers.
**Success/Failure condition:** $np_0 = (500)(0.25) = 125$ and $nq_0 = (500)(0.75) = 375$ are both greater than 10, so the sample is large enough.

The conditions have been satisfied, so a Normal model can be used to model the sampling distribution of the proportion, with $\mu_{\hat{p}} = p_0 = 0.25$ and

$$SD(\hat{p}) = \sqrt{\frac{p_0 q_0}{n}} = \sqrt{\frac{(0.25)(0.75)}{500}} \approx 0.0194 .$$

We can perform a one-proportion z-test. The observed proportion of interested readers is $\hat{p} = \dfrac{137}{500} = 0.274$.

Since the *P*-value = 0.1076 is high, we fail to reject the null hypothesis. There is little evidence to suggest that the proportion of interested readers is

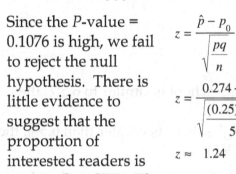

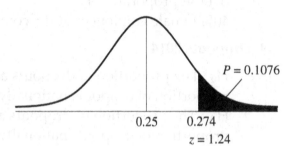

greater than 25%. The magazine should not publish the online edition.

## 37. Women executives.

$H_0$ : The proportion of female executives is similar to the overall proportion of female employees at the company. ($p = 0.40$)
$H_A$ : The proportion of female executives is lower than the overall proportion of female employees at the company. ($p < 0.40$)

**Independence assumption:** It is reasonable to think that executives at this company were chosen independently.
**Randomization condition:** The executives were not chosen randomly, but it is reasonable to think of these executives as representative of all potential executives over many years.
**10% condition:** 43 executives are less than 10% of all executives at the company.
**Success/Failure condition:** $np_0 = (43)(0.40) = 17.2$ and $nq_0 = (43)(0.60) = 25.8$ are both greater than 10, so the sample is large enough.

The conditions have been satisfied, so a Normal model can be used to model the sampling distribution of the proportion, with $\mu_{\hat{p}} = p_0 = 0.40$ and

$$SD(\hat{p}) = \sqrt{\frac{p_0 q_0}{n}} = \sqrt{\frac{(0.40)(0.60)}{43}} \approx 0.0747 .$$

Perform a one-proportion z-test. The observed proportion is $\hat{p} = \dfrac{13}{43} \approx 0.302$.

Since the P-value = 0.0955 is high, we fail to reject the null hypothesis. There is little evidence to suggest proportion of female executives is any different from the overall proportion of 40% female employees at the company.

$$z = \frac{\hat{p} - p_0}{\sqrt{\dfrac{pq}{n}}}$$

$$z = \frac{0.302 - 0.40}{\sqrt{\dfrac{(0.40)(0.60)}{43}}}$$

$$z \approx -1.31$$

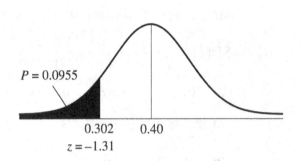

### 39. Dropouts 2014.

$H_0$ : The proportion of dropouts at this high school is similar to 6.5%, the proportion of dropouts nationally. ($p = 0.065$)
$H_A$ : The proportion of dropouts at this high school is greater than 6.5%, the proportion of dropouts nationally. ($p > 0.065$)

**Independence assumption /Randomization condition:** Assume that the students at this high school are representative of all students nationally. This is really what we are testing. The dropout rate at this high school has traditionally been close to the national rate. If we reject the null hypothesis, we will have evidence that the dropout rate at this high school is no longer close to the national rate.

**10% condition:** 1782 students are less than 10% of all students nationally.

**Success/Failure condition:** $np_0 = (1782)(0.065) = 115.83$ and $nq_0 = (1782)(0.935) = 1666.17$ are both greater than 10, so the sample is large enough.

The conditions have been satisfied, so a Normal model can be used to model the sampling distribution of the proportion, $\mu_{\hat{p}} = p_0 = 0.065$ and

$$SD(\hat{p}) = \sqrt{\frac{p_0 q_0}{n}} = \sqrt{\frac{(0.065)(0.935)}{1782}} \approx 0.00584 \, .$$

We can perform a one-proportion z-test. The observed proportion of dropouts is $\hat{p} = \dfrac{130}{1782} \approx 0.0729517$ .

Since the P-value = 0.0866 is not low, we fail to reject the null hypothesis. There is little evidence that the dropout rate at this high school is significantly higher than 6.5%.

$$z = \frac{\hat{p} - p_0}{\sqrt{\dfrac{pq}{n}}}$$

$$z = \frac{0.0729517 - 0.065}{\sqrt{\dfrac{(0.065)(0.935)}{1782}}}$$

$$z \approx 1.362$$

$P = 0.0866$

0.065   0.0729517

$z = 1.362$

**41.** **Lost luggage.**

$H_0$ : The proportion of lost luggage returned the next day is 90%. ($p = 0.90$)
$H_A$ : The proportion of lost luggage returned is lower than 90%. ($p < 0.90$)

**Independence assumption:** It is reasonable to think that the people surveyed were independent with regards to their luggage woes.
**Randomization condition:** Although not stated, we will hope that the survey was conducted randomly, or at least that these air travelers are representative of all air travelers for that airline.
**10% condition:** 122 air travelers are less than 10% of all air travelers on the airline.
**Success/Failure condition:** $np_0 = (122)(0.90) = 109.8$ and $nq_0 = (122)(0.10) = 12.2$ are both greater than 10, so the sample is large enough.

The conditions have been satisfied, so a Normal model can be used to model the sampling distribution of the proportion, with $\mu_{\hat{p}} = p_0 = 0.90$ and

$$SD(\hat{p}) = \sqrt{\frac{p_0 q_0}{n}} = \sqrt{\frac{(0.90)(0.10)}{122}} \approx 0.0272 \,.$$

We can perform a one-proportion $z$-test. The observed proportion of lost luggage is $\hat{p} = \dfrac{103}{122} \approx 0.844 \,.$

Since the $P$-value = 0.0201 is low, we reject the null hypothesis. There is evidence that the proportion of lost luggage returned the next day is lower than the 90% claimed by the airline.

$$z = \frac{\hat{p} - p_0}{\sqrt{\dfrac{pq}{n}}}$$

$$z = \frac{0.844 - 0.90}{\sqrt{\dfrac{(0.90)(0.10)}{122}}}$$

$$z \approx -2.05$$

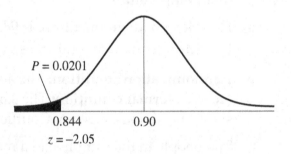

$P = 0.0201$

$0.844 \qquad 0.90$

$z = -2.05$

**43. John Wayne.**

**a)** $H_0$ : The death rate from cancer for people working on the film was similar to that predicted by cancer experts, 30 out of 220.
$H_A$ : The death rate from cancer for people working on the film was higher than the rate predicted by cancer experts.

The conditions for inference are not met, since this is not a random sample. We will assume that the cancer rates for people working on the film are similar to those predicted by the cancer experts, and a Normal model can be used to model the sampling distribution of the rate, with $\mu_{\hat{p}} = p_0 = 30/220$ and

$$SD(\hat{p}) = \sqrt{\frac{p_0 q_0}{n}} = \sqrt{\frac{\left(\frac{30}{220}\right)\left(\frac{190}{220}\right)}{220}} \approx 0.0231 \,.$$

We can perform a one-proportion $z$-test. The observed cancer rate is

$$\hat{p} = \frac{46}{220} \approx 0.209 \,.$$

$$z = \frac{\hat{p} - p_0}{SD(\hat{p})}$$

Since the $P$-value $= 0.0008$ is very low, we reject the null hypothesis. There is strong evidence that the cancer rate is higher than expected among the workers on the film.

$$z = \frac{\frac{46}{220} - \frac{30}{220}}{\sqrt{\frac{\left(\frac{30}{220}\right)\left(\frac{190}{220}\right)}{220}}}$$

$$z = 3.14$$

**b)** This does not prove that exposure to radiation may increase the risk of cancer. This group of people may be atypical for reasons that have nothing to do with the radiation.

## 45. Normal temperature.

**a)** $H_0$ : Mean body temperature is 98.6°F, as commonly assumed. $\mu = 98.6°F$
   $H_A$ : Mean body temperature is not 98.6°F. $\mu \neq 98.6°F$

**b)** **Randomization condition:** The adults were randomly selected.
   **Nearly Normal condition:** The sample of 52 adults is large, and the histogram shows no serious skewness, outliers, or multiple modes.

The people in the sample had a mean temperature of 98.285° and a standard deviation in temperature of 0.6824°. Since the conditions are satisfied, the sampling distribution of the mean can be modeled by a Student's $t$ model, with $52 - 1 = 51$ degrees of freedom, $t_{51}\left(98.6, \frac{0.6824}{\sqrt{52}}\right)$. We will use a one-sample $t$-test.

c) Since the P-value = 0.0016 is low, we reject the null hypothesis. There is strong evidence that the true mean body temperature of adults is not 98.6°F. This sample would suggest that it is significantly lower.

$$t = \frac{\bar{y} - \mu_0}{SE(\bar{y})}$$

$$t = \frac{98.285 - 98.6}{\frac{0.6824}{\sqrt{52}}}$$

$$t \approx -3.33$$

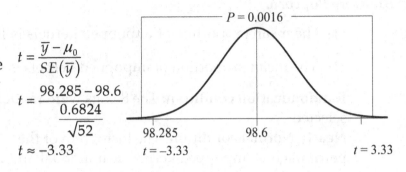

$t = -3.33$

## 47. Pizza again.

If the mean cholesterol level really does not exceed the value considered to indicate a health risk, there is a 7% probability that a random sample of this size would have a mean as high as (or higher than) the mean in this sample.

## 49. More Ruffles.

a) **Randomization condition:** The 6 bags were not selected at random, but it is reasonable to think that these bags are representative of all bags of chips.
**Nearly Normal condition:** The histogram of the weights of chips in the sample is nearly normal.

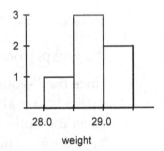

b) $\bar{y} \approx 28.78$ grams, $s \approx 0.40$ grams

c) $H_0$ : The mean weight of Ruffles bags is 28.3 grams. $\mu = 28.3$
$H_A$ : The mean weight of Ruffles bags is not 28.3 grams. $\mu \neq 28.3$

Since the conditions for inference have been satisfied, the sampling distribution of the mean can be modeled by a Student's $t$ model, with $6 - 1 = 5$ degrees of freedom, $t_5\left(28.3, \frac{0.40}{\sqrt{6}}\right)$. We will use a one-sample $t$-test.

Since the P-value = 0.032 is low, we reject the null hypothesis. There is convincing evidence that the true mean weight of bags of Ruffles potato chips is not 28.3 grams. This sample suggests that the true mean weight is significantly higher.

$$t = \frac{\bar{y} - \mu_0}{SE(\bar{y})}$$

$$t = \frac{28.78 - 28.3}{\frac{0.40}{\sqrt{6}}}$$

$$t \approx 2.94$$

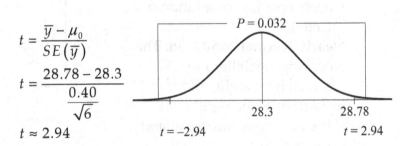

$t = 2.94$

### 51. More Popcorn.

$H_0$: The mean proportion of unpopped kernels is 10%. $(\mu = 10)$

$H_A$: The mean proportion of unpopped kernels is lower than 10%. $(\mu < 10)$

**Randomization condition:** The 8 bags were randomly selected.
**Nearly Normal condition:** The histogram of the percentage of unpopped kernels is unimodal and roughly symmetric.

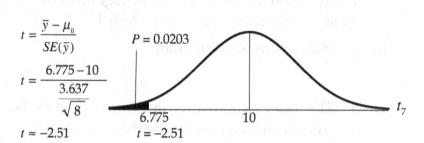

The bags in the sample had a mean percentage of unpopped kernels of 6.775 percent and a standard deviation in percentage of unpopped kernels of 3.637 percent. Since the conditions for inference are satisfied, we can model the sampling distribution of the mean percentage of unpopped kernels with a Student's $t$ model, with $8 - 1 = 7$ degrees of freedom, $t_7\left(10, \dfrac{3.637}{\sqrt{8}}\right)$.

We will perform a one-sample $t$-test.

Since the $P$-value = 0.0203 is low, we reject the null hypothesis. There is evidence to suggest the mean percentage of unpopped kernels is less than 10% at this setting.

$$t = \frac{\bar{y} - \mu_0}{SE(\bar{y})}$$

$$t = \frac{6.775 - 10}{\dfrac{3.637}{\sqrt{8}}}$$

$$t \approx -2.51$$

$P = 0.0203$

$t = -2.51$

### 53. Chips Ahoy! again.

a) **Randomization condition:** The bags of cookies were randomly selected.
**Nearly Normal condition:** The Normal probability plot is reasonably straight, and the histogram of the number of chips per bag is unimodal and symmetric.

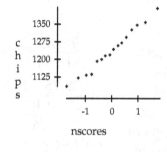

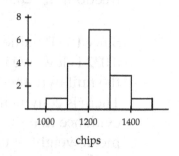

The bags in the sample had a mean number of chips of 1238.19, and a standard deviation of 94.282 chips. . Since the conditions for inference are satisfied, we can model the sampling distribution of the mean number of chips with a Student's *t* model, with 16 – 1 = 15 degrees of freedom, $t_{15}\left(1000,\dfrac{94.282}{\sqrt{16}}\right)$. We will perform a one-sample *t*-test.

**b)** $H_0$: The mean number of chips per bag is 1000. $(\mu = 1000)$

$H_A$: The mean number of chips per bag is greater than 1000. $(\mu > 1000)$

Since the P-value < 0.0001 is low, we reject the null hypothesis. There is convincing evidence that the true mean number of chips in bags of Chips Ahoy! Cookies is greater than 1000.

$$t = \frac{\bar{y} - \mu_0}{SE(\bar{y})}$$

$$t = \frac{1238.19 - 1000}{\dfrac{94.282}{\sqrt{16}}}$$

$$t \approx 10.1$$

However, since the "1000 Chip Challenge" is about individual bags, not means, the claim made by Nabisco may not be true. They claim that *all* bags of cookies have at least 1000 chips. This test doesn't really answer that question. (If the mean was around 1188 chips, the low end of our confidence interval from the question in the last chapter, and the standard deviation of the population was about 94 chips, our best estimate obtained from our sample, a bag containing 1000 chips would be about 2 standard deviations below the mean. This is not likely to happen, but not an outrageous occurrence. These data do not provide evidence that the "1000 Chip Challenge" is true).

**55. Maze.**

**a) Independence assumption:** It is reasonable to think that the rats' times will be independent, as long as the times are for different rats.

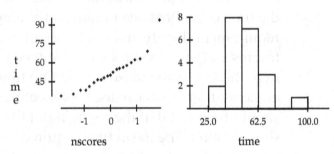

**Nearly Normal condition:** There is an outlier in both the Normal probability plot and the histogram that should probably be eliminated before continuing the test. One rat took a long time to complete the maze.

**b)** $H_0$: The mean time for rats to complete this maze is 60 seconds. $(\mu = 60)$

$H_A$: The mean time for rats to complete this maze is at most 60 seconds. $(\mu < 60)$

The rats in the sample finished the maze with a mean time of 52.21 seconds and a standard deviation in times of 13.5646 seconds. We will model the sampling distribution of the mean time in which rats complete the maze with a Student's $t$ model, with

$21 - 1 = 20$ degrees of freedom, $t_{20}\left(60, \dfrac{13.5646}{\sqrt{21}}\right)$. We will perform a one-sample

$t$-test.

Since the $P$-value = 0.008 is low, we reject the null hypothesis. There is evidence that the mean time required for rats to finish the maze is less than 60 seconds.

$$t = \frac{\overline{y} - \mu_0}{SE\left(\overline{y}\right)}$$

$$t = \frac{52.21 - 60}{\dfrac{13.5646}{\sqrt{21}}}$$

$$t \approx -2.63$$

$P = 0.008$

$52.21 \qquad 60$

$t = -2.63$

Without the outlier, the rats in the sample finished the maze with a mean time of 50.13 seconds and standard deviation in times of 9.90 seconds. Since the conditions for inference are satisfied, we can model the sampling distribution of the mean time in which rats complete the maze with a Student's $t$ model, with

$20 - 1 = 19$ degrees of freedom, $t_{19}\left(60, \dfrac{9.90407}{\sqrt{20}}\right)$. We will use a one-sample $t$-test.

This test results in a value of $t = -4.46$, and a one-sided $P$-value = 0.0001. Since the $P$-value is low, we reject the null hypothesis. There is evidence that the mean time required for rats to finish the maze is less than 60 seconds.

According to both tests, there is evidence that the mean time required for rats to complete the maze is less than 60 seconds. This maze meets the requirement that the time be at most one minute, on average. The original question had a requirement that the maze take about one minute. If we performed a two-tailed test instead, the P-values would still be very small, and we would reject the null hypothesis that the maze completion time was one minute, on average. There is evidence that the maze does not meet the "one-minute average" requirement. It should be noted that the test without the outlier is the appropriate test. The one slow rat made the mean time required seem much higher than it probably was.

**57. Maze revisited.**

a) To find the 99% bootstrap confidence interval, use the 0.5th percentile and the 99.5th percentile. We are 99% confident that the true mean maze completion time is between 45.271 and 59.939 seconds.

b) The sampling distribution is slightly skewed to the right because of the outlier at 93.8 seconds.

c) Because of the outlier, the confidence interval is not symmetric around the sample mean of 52.21 seconds.

d) Because 60 seconds is not in the interval, it is not plausible that the mean time is 60 seconds. The mean time appears to be less than 60 seconds.

e) We want the proportion of cases farther than 7.79 seconds from 60 seconds. On the left, that is fewer than 0.05% of cases. On the right, it is fewer than 0.5% of cases. We might estimate the P-value at about 0.004.

**59. Cholesterol.**

a) The mean total cholesterol is 234.701 mg/dL

b) (Answers may vary slightly, depending on your bootstrap results.) We are 95% that the true mean total cholesterol level of adults is between 232.35 and 237.26 mg/dL.

c) (Answers may vary slightly, depending on your bootstrap results.) In our bootstrapped test, bootstrapped means as low as the sample mean of 234.701 never occurred. The bootstrapped P-value is < 0.001 using 1000 samples.

d) Both the interval and the test indicate that a mean of 240 mg/dL is not plausible. The mean cholesterol level of the Framingham participants is lower than 240 mg/dL by an amount too large to be attributed to random variability.

# Chapter 16 – More About Tests and Intervals

**Section 16.1**

1. **a)** False. It provides evidence against it but does not show it is false.

   **b)** False. The P-value is not the probability that the null hypothesis is true.

   **c)** True

   **d)** False. Whether a P-value provides enough evidence to reject the null hypothesis depends on the risk of a Type I error that one is willing to assume (the $\alpha$ level).

3. **P-values.**

   **a)** False. A low *P*-value provides evidence for rejecting the null hypothesis.

   **b)** False. It results in rejecting the null hypothesis, but does not prove that it is false.

   **c)** False. A high *P*-value shows that the data are consistent with the null hypothesis but does not prove that the null hypothesis is true.

   **d)** False. Whether a *P*-value provides enough evidence to reject the null hypothesis depends on the risk of a type I error that one is willing to assume (the $\alpha$ level).

**Section 16.2**

5. **Alpha true and false.**

   **a)** True.

   **b)** False. The alpha level is set independently and does not depend on the sample size.

   **c)** False. The *P*-value would have to be less than 0.01 to reject the null hypothesis.

   **d)** False. It simply means we do not have enough evidence at that alpha level to reject the null hypothesis.

7. **Critical values.**

   **a)** $z = \pm 1.96$

   **b)** $z = 1.645$

   **c)** $t = \pm 2.03$

   **d)** $z = 2.33$ (*n* is not relevant for critical values of *z*)

   **e)** $z = -2.33$

**Section 16.3**

### 9. Significant?

a) If 98% of children have really been vaccinated, there is practically no chance of observing 97.4% of children (in a sample of 13,000) vaccinated by natural sampling variation alone.

b) We conclude that the proportion of children who have been vaccinated is below 98%, but a 95% confidence interval would show that the true proportion is between 97.1% and 97.7%. Most likely a decrease from 98% to 97.7% would not be considered important. The 98% figure was probably an approximate figure anyway. However, if the 98% figure was not as estimate, and with 1,000,000 kids per year vaccinated, even 0.1% represents 1,000 kids, so this may be important.

**Section 16.4**

### 11. Errors.

a) Type I error. The actual value is not greater than 0.3 but they rejected the null hypothesis.

b) No error. The actual value is 0.50, which was not rejected.

c) Type II error. The actual value was 55.3 points, which is greater than 52.5.

d) Type II error. The null hypothesis was not rejected, but it was false. The true relief rate was greater than 0.25.

**Chapter Exercises**

### 13. P-value.

If the effectiveness of the new poison ivy treatment is the same as the effectiveness of the old treatment, the chance of observing an effectiveness this large or larger in a sample of the same size is 4.7% by natural sampling variation alone.

### 15. Alpha.

Since the null hypothesis was rejected at $\alpha = 0.05$, the *P*-value for the researcher's test must have been less than 0.05. He would have made the same decision at $\alpha = 0.10$, since the *P*-value must also be less than 0.10. We can't be certain whether or not he would have made the same decision at $\alpha = 0.01$, since we only know that the *P*-value was less than 0.05. It may have been less than 0.01, but we can't be sure.

## 17. Groceries.

a) **Randomization condition:** We will assume that the Yahoo survey was conducted randomly.
**10% condition:** 2400 is less than 10% of all men.
**Success/Failure condition:** $n\hat{p} = 1224$ and $n\hat{q} = 1176$ are both greater than 10, so the sample is large enough.

Since the conditions are met, we can use a one-proportion z-interval to estimate the percentage of men who identify themselves as the primary grocery shopper in their household.

$$\hat{p} \pm z^* \sqrt{\frac{\hat{p}\hat{q}}{n}} = \left(\frac{1224}{2400}\right) \pm 2.326 \sqrt{\frac{\left(\frac{1224}{2400}\right)\left(\frac{1176}{2400}\right)}{2400}} = (48.6\%, 53.4\%)$$

We are 98% confident that between 48.6% and 53.4% of all men identify themselves as the primary grocery shopper in their household.

b) Since 45% is not in the interval, there is strong evidence that more than 45% of all men identify themselves as the primary grocery shopper in their household.

c) The significance level of this test is $\alpha = 0.01$. It's an upper tail test based on a 98% confidence interval.

## 19. Approval 2016.

a) **Randomization condition:** The adults were randomly selected.
**10% condition:** 1500 adults represent less than 10% of all adults.
**Success/Failure condition:** $n\hat{p} = (1500)(0.57) = 855$ and $n\hat{q} = (1500)(0.43) = 645$ are both greater than 10, so the sample is large enough.

Since the conditions are met, we can use a one-proportion z-interval to estimate Barack Obama's approval rating.

$$\hat{p} \pm z^* \sqrt{\frac{\hat{p}\hat{q}}{n}} = (0.57) \pm 1.960 \sqrt{\frac{(0.57)(0.43)}{1500}} = (0.545, 0.595)$$

We are 95% confident that Barack Obama's approval rating is between 54.5% and 59.5%.

b) Since 52% is not within the interval, this is not a plausible value for Barack Obama's approval rating. There is evidence against the null hypothesis.

## 21. Dogs.

a) We cannot construct a confidence interval for the rate of occurrence of early hip dysplasia among 6-month old puppies because only 5 of 42 puppies were found with early hip dysplasia. The Success/Failure condition is not satisfied.

b) We could not use a bootstrap confidence interval. The sample size is too small.

**23. Loans.**

a) The bank has made a Type II error. The person was not a good credit risk, and the bank failed to notice this.

b) The bank has made a Type I error. The person was a good credit risk, and the bank was convinced that he/she was not.

c) By making it easier to get a loan, the bank has reduced the alpha level. It takes less evidence to grant the person the loan.

d) The risk of Type I error is decreased and the risk of Type II error has increased.

**25. Second loan.**

a) Power is the probability that the bank denies a loan that could not have been repaid.

b) To increase power, the bank could raise the cutoff score.

c) If the bank raised the cutoff score, a larger number of trustworthy people would be denied credit, and the bank would lose the opportunity to collect the interest on these loans.

**27. Homeowners 2015.**

a) The null hypothesis is that the level of home ownership does not rise. The alternative hypothesis is that it rises.

b) In this context, a Type I error is when the city concludes that home ownership is on the rise, but in fact, the tax breaks don't help.

c) In this context, a Type II error is when the city abandons the tax breaks, thinking they don't help, when in fact they were helping.

d) A Type I error causes the city to forego tax revenue, while a Type II error withdraws help from those who might have otherwise been able to buy a house.

e) The power of the test is the city's ability to detect an actual increase in home ownership.

**29. Testing cars.**

$H_0$ : The shop is meeting the emissions standards.
$H_A$ : The shop is not meeting the emissions standards.

a) Type I error is when the regulators decide that the shop is not meeting standards when they actually are meeting the standards.

b) Type II error is when the regulators certify the shop when they are not meeting the standards.

c) Type I would be more serious to the shop owners. They would lose their certification, even though they are meeting the standards.

**d)** Type II would be more serious to environmentalists. Shops are allowed to operate, even though they are allowing polluting cars to operate.

## 31. Cars again.

**a)** The power of the test is the probability of detecting that the shop is not meeting standards when they are not.

**b)** The power of the test will be greater when 40 cars are tested. A larger sample size increases the power of the test.

**c)** The power of the test will be greater when the level of significance is 10%. There is a greater chance that the null hypothesis will be rejected.

**d)** The power of the test will be greater when the shop is out of compliance "a lot". Larger problems are easier to detect.

## 33. Equal opportunity?

$H_0$ : The company is not discriminating against minorities.
$H_A$ : The company is discriminating against minorities.

**a)** This is a one-tailed test. They wouldn't sue if "too many" minorities were hired.

**b)** Type I error would be deciding that the company is discriminating against minorities when they are not discriminating.

**c)** Type II error would be deciding that the company is not discriminating against minorities when they actually are discriminating.

**d)** The power of the test is the probability that discrimination is detected when it is actually occurring.

**e)** The power of the test will increase when the level of significance is increased from 0.01 to 0.05.

**f)** The power of the test is lower when the lawsuit is based on 37 employees instead of 87. Lower sample size leads to less power.

## 35. Software for learning.

**a)** The test is one-tailed. We are testing to see if an increase in average score is associated with the software.

**b)** $H_0$ : The average score does not change following the use of software. ($\mu = 105$)
$H_A$ : The average score increases following the use of the software. ($\mu > 105$)

**c)** The professor makes a Type I error if he buys the software when the average score has not actually increased.

**d)** The professor makes a Type II error if he doesn't buy the software when the average has actually increased.

e) The power of the test is the probability of buying the software when the average score has actually increased.

## 37. Software, part II.

a) $H_0$ : The average score does not change following the use of software. ($\mu = 105$)

$H_A$ : The average score increases following the use of the software. ($\mu > 105$)

**Randomization condition:** This year's class of 203 students is probably representative of all stats students.

**Nearly Normal condition:** We don't have the scores from the 203 individuals, so we can't check a plot of the data. However, with a sample this large, the Central Limit Theorem allows us to model the sampling distribution of the means with a *t*-distribution.

The mean score was 108 points, with a standard deviation of 8.7 points. Since the conditions for inference are satisfied, we can model the sampling distribution of the mean score with a Student's *t* model, with 203 – 1 = 202 degrees of freedom,

$t_{202}\left(108, \dfrac{8.7}{\sqrt{203}}\right)$. We will perform a one-sample *t*-test.

The value of *t* is approximately 4.91, which results in a *P*-value of less than 0.0001, so we reject the null hypothesis. There is strong evidence that the mean score has increased since use of the software program was implemented. As long as the professor feels confident that this class of stats students is representative of all potential students, then he should buy the program.

$$t = \frac{\bar{y} - \mu_0}{\dfrac{\sigma}{\sqrt{n}}}$$

$$t = \frac{108 - 105}{\dfrac{8.7}{\sqrt{203}}}$$

$$t = 4.91$$

If you used a 95% confidence interval to assess the effectiveness of the program: $\bar{y} \pm t^*_{n-1}\left(\dfrac{s}{\sqrt{n}}\right) = 108 \pm t^*_{202}\left(\dfrac{8.7}{\sqrt{203}}\right) \approx (106.8,\ 109.2)$

We are 95% confident that the mean score is between 106.8 and 109.2. Since 105 is above the interval, this provides evidence that the mean score has increased following the implementation of the software program.

b) The mean score on the exam only increased by 1 to 4 points. This small difference might not be enough to be worth the cost of the program.

## 39. TV safety.

a) This is an upper-tail test. We hope to show that the TV stand will hold 500 or more pounds easily.

b) The inspectors will commit a Type I error if they decide the TV stands are safe when they are not.

c) The inspectors will commit a Type II error if they decide the TVs are unsafe when they are actually safe.

**41. TV safety, revisited.**

a) To decrease the likelihood of producing an unsafe TV stand, they should decrease $\alpha$. This lower the chance of making a Type I error.

b) The power of the test is the ability of the inspectors to determine that the TV stand is safe when it is actually capable of holding more than 500 pounds.

c) The company could increase the power of the test by lowering the standard deviation by testing more stands. This could prove costly and would require more time to test. They could also increase $\alpha$, but then they will commit Type I errors with greater frequency, approving stands that cannot hold 500 pounds of more. Finally, they could require that TV stands have a higher weight capacity than 500 pounds as the standard. Again, that might prove costly, since they would be rejecting many more stands that were safe.

**43. Two coins.**

a) The alternative hypothesis is that your coin produces 30% heads.

b) Reject the null hypothesis if the coin comes up tails. Otherwise, fail to reject.

c) There is a 10% chance that the coin comes up tails if the null hypothesis is true, so alpha is 10%.

d) Power is our ability to detect the 30% coin. That coin will come up tails 70% of the time. That's the power of our test.

e) To increase the power and lower the probability of Type I error at the same time, simply flip the coin more times.

**45. Hoops.**

$H_0$ : The player's foul-shot percentage is only 60%. ($p = 0.60$)
$H_A$ : The player's foul-shot percentage is better than 60%. ($p > 0.60$)

a) There are only two possible outcomes, make the shot and miss the shot. The probability of making any shot is constant at $p = 0.60$. Assume that the shots are independent of each other.

There are 10 different ways to make 9 out of 10. To look at it another way, there are 10 ways to choose the shot to miss.

$$P(\text{makes at least 9 out of 10}) = P(\text{makes 9}) + P(\text{makes 10})$$
$$= 10(0.60)^9 (0.40)^1 + (0.60)^{10}$$
$$\approx 0.0464$$

b) The coach made a Type I error.

**c)** The power of the test to detect improvement in foul-shooting can be increased by increasing the number of shots, or by keeping the number of shots at 10 but increasing the level of significance by declaring that 8, 9, or 10 shots made will convince the coach that the player has improved. In other words, the coach can increase the power of the test by lowering the standard of proof.

**47. Chips Ahoy! bootstrapped.**

The sample mean of the original data set was 1238.19 chips. If the null hypothesis were true, and the true mean were 1000 chips, we didn't see means that extreme or more extreme in any of our 10,000 bootstrap samples. That's less than 1/10,000 times, or a P-value < 0.00001.

## Review of Part IV – From the Data at Hand to the World at Large

**R4.1.  Quality Control.**

a)  $P(\text{defect}) = P(\text{cosm.}) + P(\text{func.}) - P(\text{cosm. and func.})$
$$= 0.29 + 0.07 - 0.02 = 0.34$$

Or, from the Venn: $0.27 + 0.02 + 0.05 = 0.34$

b)  $P(\text{cosm. and no func.})$

$= P(\text{cosm.}) - P(\text{cosm. and func.})$
$= 0.29 - 0.02 = 0.27$

Or, from the Venn: 0.27 (region inside Cosmetic circle, outside Functional circle)

c)  $P(\text{func.} \mid \text{cosm.}) = \dfrac{P(\text{func. and cosm.})}{P(\text{cosm.})} = \dfrac{0.02}{0.29} \approx 0.069$

From the Venn, consider only the region inside the Cosmetic circle.  The probability that the car has a functional defect is 0.02 out of a total of 0.29 (the entire Cosmetic circle).

d)  The two kinds of defects are not disjoint events, since 2% of cars have both kinds.

e)  Approximately 6.9% of cars with cosmetic defects also have functional defects.  Overall, the probability that a car has a cosmetic defect is 7%.  The probabilities are estimates, so these are probably close enough to say that they two types of defects are independent.

**R4.3.  Emergency switch.**

Construct a Venn diagram of the disjoint outcomes.

a)  From the Venn diagram, 3% of the workers were unable to operate the switch with either hand.

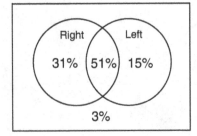

b)  $P(\text{left} \mid \text{right}) = \dfrac{P(\text{left and right})}{P(\text{right})} = \dfrac{0.51}{0.82} \approx 0.622$

About 62% of the workers who could operate the switch with their right hands could also operate it with left hands.  Overall, the probability that a worker could operate the switch with his right hand was 66%.  Workers who could operate the switch with their right hands were less likely to be able to operate the switch with their left hand, so success is not independent of hand.

c)  Success with right and left hands are not disjoint events.  51% of the workers had success with both hands.

**R4.5. Leaky gas tanks.**

a) $H_0$: The proportion of leaky gas tanks is 77%. ($p = 0.77$)
   $H_A$: The proportion of leaky gas tanks is less than 77%. ($p < 0.77$)

b) **Randomization condition:** A random sample of 47 service stations in California was taken.
   **10% condition:** 47 stations are less than 10% of all service stations in California.
   **Success/Failure condition:** $np = (47)(0.77) = 36.19$ and $nq = (47)(0.23) = 10.81$ are both greater than 10, so the sample is large enough.

c) Since the conditions have been satisfied, a Normal model can be used to model the sampling distribution of the proportion, with $\mu_{\hat{p}} = p = 0.77$ and

$$\sigma(\hat{p}) = \sqrt{\frac{pq}{n}} = \sqrt{\frac{(0.77)(0.23)}{47}} \approx 0.06138 \,.$$ We can perform a one-proportion $z$-test.

The observed proportion of leaky gas tanks is $\hat{p} = \dfrac{33}{47} \approx 0.7021$.

d) Since the $P$-value = 0.134 is relatively high, we fail to reject the null hypothesis. There is little evidence that the proportion of leaky gas tanks in

$$z = \frac{\hat{p} - p_0}{\sqrt{\dfrac{p_0 q_0}{n}}}$$

$$z = \frac{0.7021 - 0.77}{\sqrt{\dfrac{(0.77)(0.23)}{47}}}$$

$$z \approx -1.106$$

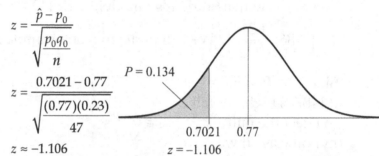

$P = 0.134$

$z = -1.106$

California is less than 77%. The new program doesn't appear to be effective in decreasing the proportion of leaky gas tanks.

e) If the program actually works, we haven't done anything *wrong*. Our methods are correct. Statistically speaking, we have committed a Type II error.

f) In order to decrease the probability of making this type of error, we could lower our standards of proof, by raising the level of significance. This will increase the power of the test to detect a decrease in the proportion of leaky gas tanks. Another way to decrease the probability that we make a Type II error is to sample more service stations. This will decrease the variation in the sample proportion, making our results more reliable.

g) Increasing the level of significance is advantageous, since it decreases the probability of making a Type II error, and increases the power of the test. However, it also increases the probability that a Type I error is made, in this case, thinking that the program is effective when it really is not effective.
   Increasing the sample size decreases the probability of making a Type II error and increases power, but can be costly and time-consuming.

**R4.7.  Babies.**

**a)** This is a one-sample test because the data from Australia are census data, not sample data.

**b)** H₀: The mean weight of newborns in the U.S. is 7.86 pounds, the same as the mean weight of Australian babies. $(\mu = 7.86)$

H_A: The mean weight of newborns in the U.S. is not the same as the mean weight of Australian babies. $(\mu \neq 7.86)$

**Randomization condition:** Assume that the babies at this Missouri hospital are representative of all U.S. newborns. (Given)
**Nearly Normal condition:** We don't have the actual data, so we cannot look at a graphical display, but since the sample is large, it is safe to proceed.

The babies in the sample had a mean weight of 7.68 pounds and a standard deviation in weight of 1.31 pounds. Since the conditions for inference are satisfied, we can model the sampling distribution of the mean weight of U.S. newborns with a Student's $t$ model, with 112 – 1 = 111 degrees of freedom,

$t_{111}\left(7.86, \dfrac{1.31}{\sqrt{112}}\right)$. We will perform a one-sample $t$-test.

Since the P-value = 0.1487 is high, we fail to reject the null hypothesis.  If we believe that the babies at this Missouri hospital are representative of all U.S. babies, there is

$$t = \frac{\bar{y} - \mu_0}{SE(\bar{y})}$$

$$t = \frac{7.68 - 7.86}{\dfrac{1.31}{\sqrt{112}}}$$

$$t \approx -1.45$$

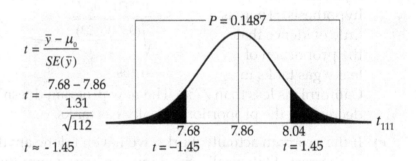

little evidence to suggest that the mean weight of U.S. babies is different than the mean weight of Australian babies.

**R4.9.  Color-blind.**

**a)** **Randomization condition:** The 325 male students are probably representative of all males.
**10% condition:** 325 male students are less than 10% of the population of males.
**Success/Failure condition:** $np = (325)(0.08) = 26$ and $nq = (325)(0.92) = 299$ are both greater than 10, so the sample is large enough.

Since the conditions have been satisfied, a Normal model can be used to model the sampling distribution of the proportion of colorblind men among 325 students.

**b)** $\mu_{\hat{p}} = p = 0.08$      $\sigma(\hat{p}) = \sqrt{\dfrac{pq}{n}} = \sqrt{\dfrac{(0.08)(0.92)}{325}} \approx 0.015$

**c)**

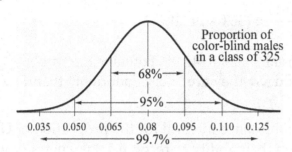

**d)** According to the Normal model, we expect about 68% of classes with 325 males to have between 6.5% and 9.5% colorblind males. We expect about 95% of such classes to have between 5% and 11% colorblind males. About 99.7% of such classes are expected to have between 3.5% and 12.5% colorblind males.

**R4.11. More twins.**

**a)** There is only one way for all five women to have no twins.

$$P(\text{none have twins}) = (0.9)^5$$
$$\approx 0.590$$

**b)** There are 5 different ways this can happen. The woman who has twins can be the first woman, or the second woman, or the third, and so on.

$$P(\text{one has twins}) = 5(0.1)^1(0.9)^4$$
$$\approx 0.328$$

**c)** This gets a bit complex. There are 10 ways equally likely ways that the 5 women could have exactly 3 sets of twins, and 5 equally likely ways that the 5 women could have exactly 4 sets of twins. There is only one way that all 5 women could have twins.

$P(\text{at least three will have twins})$

$= P(\text{exactly three}) + P(\text{exactly four}) + P(\text{exactly five})$

$= 10(0.1)^3(0.9)^2 + 5(0.1)^4(0.9)^1 + (0.1)^5(0.9)^0$

$= 0.00856$

**R4.13. Fake news.**

**a)** Pew believes the true proportion of all American adults who think that fake news is causing confusion is within 3.6 percentage points of the estimated 64% — namely, between 60.4% and 67.6%.

**b)** **Randomization condition:** The 1002 U.S. adults were sampled randomly. **10% condition:** 1002 adults are less than 10% of the population all U.S. adults. **Success/Failure condition:** $n\hat{p} = (1002)(0.39) = 391$ and $n\hat{q} = (1002)(0.92) = 611$ are both greater than 10, so the sample is large enough.

Since the conditions have been satisfied, a Normal model can be used to model the sampling distribution of the proportion of U.S. adults who say they are "very confident" that they can recognize fake news.

$$\hat{p} \pm z^* \sqrt{\frac{\hat{p}\hat{q}}{n}} = (0.39) \pm 1.960 \sqrt{\frac{(0.39)(0.61)}{1002}} = (36.0\%, 42.0\%)$$

We are 95% confident that the interval 36.0% to 42.0% contains the true proportion of U.S. adults who would say they are "very confident" that they can recognize fake news.

c) At the same level of confidence, a confidence interval for the proportion of U.S. adults who are not at all confident in their ability to recognize fake news would be narrower than the confidence interval for the proportion of U.S. adults who are very confident in their ability to do so. The standard error for proportions farther from 0.5 is smaller than the standard error for proportions closer to 0.5. We calculated the width of the interval in the previous part to be 6 percentage points.

$$\hat{p} \pm z^* \sqrt{\frac{\hat{p}\hat{q}}{n}} = (0.06) \pm 1.960 \sqrt{\frac{(0.06)(0.94)}{1002}} = (4.5\%, 7.5\%)$$

We are 95% confident that the interval 4.5% to 7.5% contains the true proportion of U.S. adults who would say they are "not at all confident" that they can recognize fake news. This interval is only 3 percentage points wide, half the width of the previous interval.

**R4.15. Passing stats.**

Organize the information in a tree diagram.

a)

$P(\text{Passing Statistics})$
   $= P(\text{Scedastic and Pass})$
      $+ P(\text{Kurtosis and Pass})$
   $\approx 0.4667 + 0.25$
   $\approx 0.717$

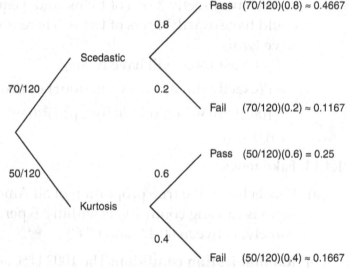

b)

$$P(\text{Kurtosis} \mid \text{Fail}) = \frac{P(\text{Kurtosis and Fail})}{P(\text{Fail})} \approx \frac{0.1667}{0.1167 + 0.1667} \approx 0.588$$

**R4.17. Stocks.**

a)  $P(\text{market will rise for 3 consecutive years}) = (0.73)^3 \approx 0.389$

b)  There are 10 different ways to choose the three years in which the market will rise, out of the five years. Each of these ways has the same basic probability.

$P(\text{market will rise in 3 out of 5 years}) = 10(0.73)^3(0.27)^2 \approx 0.284$

c)  $P(\text{fall in at least 1 of next 5 years}) = 1 - P(\text{no fall in 5 years}) = 1 - (0.73)^5 \approx 0.793$

d)  This gets a bit complex. There are 210 ways to choose the 6 years in which the market will rise out of 10, 120 ways to choose the 7 of 10, 45 ways to choose the 8 of 10, 10 ways to choose the 9 of 10, and 1 way to choose the 1 of 10.
    $P(\text{rises in the majority of years in a decade})$

$$= 210(0.73)^6(0.27)^4 + 120(0.73)^7(0.27)^3 + 45(0.73)^8(0.27)^2 + 9(0.73)^9(0.27)^1 + (0.73)^{10}$$

$$\approx 0.896$$

**R4.19. Gay marriage 2016.**

a)  **Randomization condition:** Pew Research randomly selected 2254 U.S. adults.
    **10% condition:** 2254 results is less than 10% of all U.S. adults.
    **Success/Failure condition:** $n\hat{p} = (2254)(0.57) = 1285$ and $n\hat{q} = (2254)(0.43) = 969$ are both greater than 10, so the sample is large enough.

Since the conditions are met, we can use a one-proportion $z$-interval to estimate the percentage of U.S. adults who support marriage equality.

$$\hat{p} \pm z^* \sqrt{\frac{\hat{p}\hat{q}}{n}} = (0.57) \pm 1.960 \sqrt{\frac{(0.57)(0.43)}{2254}} = (55.0\%, 59.0\%)$$

We are 95% confident that between 55.0% and 59.0% of U.S. adults support marriage equality. (If you use technology, and counts of 1285 of 2254 in support, you will get the slightly different interval 55.0% to 59.1%.)

b)  Since the interval is entirely above 50%, there is evidence that a majority of U.S. adults support marriage equality.

c)

$$ME = z^* \sqrt{\frac{\hat{p}\hat{q}}{n}}$$

$$0.02 = 2.326 \sqrt{\frac{(0.50)(0.50)}{n}}$$

$$n = \frac{(2.326)^2(0.50)(0.50)}{(0.02)^2}$$

$$n \approx 3382 \text{ people}$$

We do not know the true proportion of U.S. adults who support marriage equality, so use $\hat{p} = \hat{q} = 0.50$, for the most cautious estimate. In order to determine the proportion of U.S. adults who support marriage equality to within 2% with 98% confidence, we would have to sample at least 3382 people. (Using $\hat{p} = 0.57$ from the sample gives a slightly lower estimate of about 3315 people.)

**R4.21. Jerseys.**

a) $P(\text{all four kids get the same color}) = 4\left(\dfrac{1}{4}\right)^4 \approx 0.0156$

(There are four different ways for this to happen, one for each color.)

b) $P(\text{all four kids get white}) = \left(\dfrac{1}{4}\right)^4 \approx 0.0039$

c) $P(\text{all four kids get white}) = \left(\dfrac{1}{6}\right)\left(\dfrac{1}{4}\right)^3 \approx 0.0026$

**R4.23. Polling disclaimer.**

a) It is not clear what specific question the pollster asked. Otherwise, they did a great job of identifying the W's.

b) A sample that was stratified by age, sex, region, and education was used.

c) The margin of error was 4%.

d) Since "no more than 1 time in 20 should chance variations in the sample cause the results to vary by more than 4 percentage points", the confidence level is $19/20 = 95\%$.

e) The subgroups had smaller sample sizes than the larger group. The standard errors in these subgroups were larger as a result, and this caused the margins of error to be larger.

f) They cautioned readers about response bias due to wording and order of the questions.

**R4.25. Largemouth bass.**

a) One would expect many small fish, with a few large fish.

b) We cannot determine the probability that a largemouth bass caught from the lake weighs over 3 pounds because we don't know the exact shape of the distribution. We know that it is NOT Normal.

c) It would be quite risky to attempt to determine whether or not the mean weight of 5 fish was over 3 pounds. With a skewed distribution, a sample of size 5 is not large enough for the Central Limit Theorem to guarantee that a Normal model is appropriate to describe the distribution of the mean.

d) A sample of 60 randomly selected fish is large enough for the Central Limit Theorem to guarantee that a Normal model is appropriate to describe the sampling distribution of the mean, as long as 60 fish is less than 10% of the population of all the fish in the lake.

The mean weight is $\mu = 3.5$ pounds, with standard deviation $\sigma = 2.2$ pounds. Since the sample size is sufficiently large, we can model the sampling distribution of the mean weight of 60 fish with a Normal model, with

$$\mu_{\bar{y}} = 3.5 \text{ pounds and standard deviation } \sigma(\bar{y}) = \frac{2.2}{\sqrt{60}} \approx 0.284 \text{ pounds}.$$

According to the Normal model, the probability that 60 randomly selected fish average more than 3 pounds is approximately 0.961.

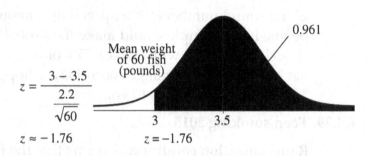

$$z = \frac{3 - 3.5}{\frac{2.2}{\sqrt{60}}}$$

$$z \approx -1.76$$

**R4.27. Language.**

a) **Randomization condition:** 60 people were selected at random.
   **10% condition:** The 60 people represent less than 10% of all people.
   **Success/Failure condition:** $np = (60)(0.80) = 48$ and $nq = (60)(0.20) = 12$ are both greater than 10.

   Therefore, the sampling distribution model for the proportion of 60 randomly selected people who have left-brain language control is Normal, with

   $$\mu_{\hat{p}} = p = 0.80 \text{ and standard deviation } \sigma(\hat{p}) = \sqrt{\frac{pq}{n}} = \sqrt{\frac{(0.80)(0.20)}{60}} \approx 0.0516.$$

b) According to the Normal model, the probability that over 75% of these 60 people have left-brain language control is approximately 0.834.

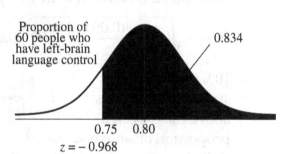

$$z = \frac{\hat{p} - \mu_{\hat{p}}}{\sqrt{\frac{pq}{n}}}$$

$$z = \frac{0.75 - 0.80}{\sqrt{\frac{(0.80)(0.20)}{60}}}$$

$$z \approx -0.968$$

c) If the sample had consisted of 100 people, the probability would have been higher. A larger sample results in a smaller standard deviation for the sample proportion.

d) Answers may vary. Let's consider three standard deviations below the expected proportion to be "almost certain". It would take a sample of (exactly!) 576 people to make sure that 75% would be 3 standard deviations below the expected percentage of people with left-brain language control.

$$z = \frac{\hat{p} - \mu_{\hat{p}}}{\sqrt{\dfrac{pq}{n}}}$$

$$-3 = \frac{0.75 - 0.80}{\sqrt{\dfrac{(0.80)(0.20)}{n}}}$$

Using round numbers for $n$ instead of $z$, about 500 people in the sample would make the probability of choosing a sample with at least 75% of the people having left-brain language control is a whopping 0.997. It all depends on what "almost certain" means to you.

$$n = \frac{(-3)^2(0.80)(0.20)}{(0.75 - 0.80)^2} = 576$$

### R4.29. Teen smoking 2015.

**Randomization condition:** Assume that the freshman class is representative of all teenagers. This may not be a reasonable assumption. There are many interlocking relationships between smoking, socioeconomic status, and college attendance. This class may not be representative of all teens with regards to smoking simply because they are in college. Be cautious with your conclusions!
**10% condition:** The freshman class is less than 10% of all teenagers.
**Success/Failure condition:** $np = (522)(0.093) = 48.546$ and $nq = (522)(0.907) = 473.454$ are both greater than 10.

Therefore, the sampling distribution model for the proportion of 522 students who smoke is Normal, with $\mu_{\hat{p}} = p = 0.093$, and standard deviation

$$\sigma(\hat{p}) = \sqrt{\frac{pq}{n}} = \sqrt{\frac{(0.093)(0.907)}{522}} \approx 0.0127 .$$

10% is about 0.551 standard deviations above the expected proportion of smokers. If the true proportion of smokers is 9.3%, the Normal model predicts that the probability that more than 10% of these students smoke is approximately 0.291.

$$z = \frac{\hat{p} - \mu_{\hat{p}}}{\sqrt{\dfrac{pq}{n}}}$$

$$z = \frac{0.1 - 0.093}{\sqrt{\dfrac{(0.093)(0.907)}{522}}}$$

$$z \approx 0.551$$

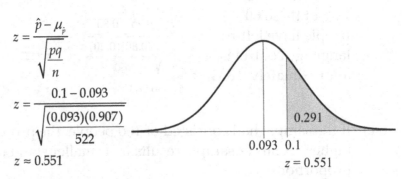

**R4.31. Alcohol abuse.**

$$ME = z^* \sqrt{\dfrac{\hat{p}\hat{q}}{n}}$$

$$0.04 = 1.645 \sqrt{\dfrac{(0.5)(0.5)}{n}}$$

$$n = \dfrac{(1.645)^2 (0.5)(0.5)}{(0.04)^2}$$

$$n \approx 423$$

The university will have to sample at least 423 students in order to estimate the proportion of students who have been drunk with in the past week to within ± 4%, with 90% confidence.

**R4.33. Pregnant?**

Organize the information in a tree diagram.

$P(\text{pregnant} \mid + \text{test})$

$$= \dfrac{P(\text{preg. and + test})}{P(+ \text{test})}$$

$$= \dfrac{0.686}{0.686 + 0.006}$$

$$\approx 0.991$$

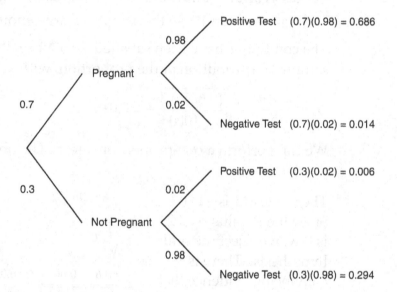

Positive Test (0.7)(0.98) = 0.686
Negative Test (0.7)(0.02) = 0.014
Positive Test (0.3)(0.02) = 0.006
Negative Test (0.3)(0.98) = 0.294

**R4.35. Fried PCs.**

a)  $H_0$: The computer is undamaged.
    $H_A$: The computer is damaged.

b)  The biggest advantage is that all of the damaged computers will be detected, since, historically, damaged computers never pass all the tests. The disadvantage is that only 80% of undamaged computers pass all the tests. The engineers will be classifying 20% of the undamaged computers as damaged.

c)  In this example, a Type I error is rejecting an undamaged computer. To allow this to happen only 5% of the time, the engineers would reject any computer that failed 3 or more tests, since 95% of the undamaged computers fail two or fewer tests.

d)  The power of the test in part c is 20%, since only 20% of the damaged machines fail 3 or more tests.

e) By declaring computers "damaged" if they fail 2 or more tests, the engineers will be rejecting only 7% of undamaged computers. From 5% to 7% is an increase of 2% in $\alpha$. Since 90% of the damaged computers fail 2 or more tests, the power of the test is now 90%, a substantial increase.

**R4.37. Approval 2016.**

$H_0$ Barack Obama's final approval rating was 66%. ($p = 0.66$)
$H_A$ Barack Obama's final approval rating was lower than 66%. ($p > 0.66$)

**Randomization condition:** The adults were chosen randomly.
**10% condition:** 1000 adults are less than 10% of all adults.
**Success/Failure condition:** $np = (1000)(0.66) = 660$ and $nq = (1000)(0.34) = 340$ are both greater than 10, so the sample is large enough.

The conditions have been satisfied, so a Normal model can be used to model the sampling distribution of the proportion, with $\mu_{\hat{p}} = p = 0.66$ and

$$\sigma(\hat{p}) = \sqrt{\frac{pq}{n}} = \sqrt{\frac{(0.66)(0.34)}{1000}} \approx 0.015.$$

We can perform a one-proportion z-test. The observed approval rating is $\hat{p} = 0.63$.

The value of $z$ is –2.00. Since the *P*-value = 0.023 is low, we reject the null hypothesis. There is convincing evidence that President Barack Obama's final approval rating was lower than the 66% approval rating of President Bill Clinton.

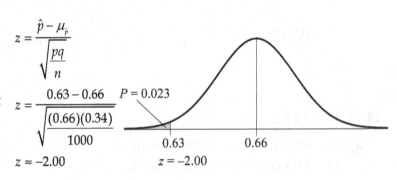

$$z = \frac{\hat{p} - \mu_{\hat{p}}}{\sqrt{\frac{pq}{n}}}$$

$$z = \frac{0.63 - 0.66}{\sqrt{\frac{(0.66)(0.34)}{1000}}}$$

$z \approx -2.00$

$P = 0.023$

**R4.39. Name recognition.**

a) The company wants evidence that the athlete's name is recognized more often than 25%.

b) Type I error means that fewer than 25% of people will recognize the athlete's name, yet the company offers the athlete an endorsement contract anyway. In this case, the company is employing an athlete that doesn't fulfill their advertising needs.

Type II error means that more than 25% of people will recognize the athlete's name, but the company doesn't offer the contract to the athlete. In this case, the company is letting go of an athlete that meets their advertising needs.

c) If the company uses a 10% level of significance, the company will hire more athletes that don't have high enough name recognition for their needs. The risk of committing a Type I error is higher.

At the same level of significance, the company is less likely to lose out on athletes with high name recognition. They will commit fewer Type II errors.

## R4.41. Dropouts.

**Randomization condition:** Assume that these subjects are representative of all anorexia nervosa patients.
**10% condition:** 198 is less than 10% of all patients.
**Success/Failure condition:** The number of dropouts, 105, and the number of subjects that remained, 93, are both greater than 10, so the samples are both large enough.

Since the conditions have been satisfied, we will find a one-proportion z-interval.

$$\hat{p} \pm z^* \sqrt{\frac{\hat{p}\hat{q}}{n}} = \left(\frac{105}{198}\right) \pm 1.960 \sqrt{\frac{\left(\frac{105}{198}\right)\left(\frac{93}{198}\right)}{198}} = (46\%, 60\%)$$

We are 95% confident that between 46% and 60% of anorexia nervosa patients will drop out of treatment programs. However, this wasn't a random sample of all patients. They were assigned to treatment programs rather than choosing their own. They may have had different experiences if they were not part of an experiment.

## R4.43. Speeding.

a) $H_0$ : The percentage of speeding tickets issued to black drivers is 16%, the same as the percentage of registered drivers who are black. ($p = 0.16$)

$H_A$ : The percentage of speeding tickets issued to black drivers is greater than 16%, the percentage of registered drivers who are black. ($p > 0.16$)

**Random condition:** Assume that this month is representative of all months with respect to the percentage of tickets issued to black drivers.
**10% condition:** 324 speeding tickets are less than 10% of all tickets.
**Success/Failure condition:** $np = (324)(0.16) = 52$ and $nq = (324)(0.84) = 272$ are both greater than 10, so the sample is large enough.

The conditions have been satisfied, so a Normal model can be used to model the sampling distribution of the proportion, with $\mu_{\hat{p}} = p = 0.16$ and

$$\sigma(\hat{p}) = \sqrt{\frac{pq}{n}} = \sqrt{\frac{(0.16)(0.84)}{324}} \approx 0.02037.$$

We can perform a one-proportion $z$-test.
The observed proportion of tickets issued is $\hat{p} = 0.25$.

Since the $P$-value $= 4.96 \times 10^{-6}$ is very low, we reject the null hypothesis. There is strong evidence that the percentage of speeding tickets issued to black drivers is greater than 16%.

$$z = \frac{\hat{p} - p_0}{SD(\hat{p})}$$

$$z \approx \frac{0.25 - 0.16}{\sqrt{\dfrac{(0.16)(0.84)}{324}}}$$

$$z \approx 4.42$$

**b)** There is strong evidence of an association between the receipt of a speeding ticket and race. Black drivers appear to be issued tickets at a higher rate than expected. However, this does not prove that racial profiling exists. There may be other factors present.

**c)** Answers may vary. The primary statistic of interest is the percentage of black motorists on this section of the New Jersey Turnpike. For example, if 80% of drivers on this section are black, then 25% of the speeding tickets being issued to black motorists is not an usually high percentage. In fact, it is probably unusually low. On the other hand, if only 3% of the motorists on this section of the turnpike are black, then there is even more evidence that racial profiling may be occurring.

**R4.45. Meals.**

$H_0$: The college student's mean daily food expense is \$10. ($\mu = 10$)

$H_A$: The college student's mean daily food expense is greater than \$10. ($\mu > 10$)

**Randomization condition:** Assume that these days are representative of all days.
**Nearly Normal condition:** The histogram of daily expenses is fairly unimodal and symmetric. It is reasonable to think that this sample came from a Normal population.

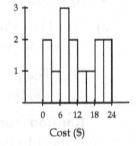

The expenses in the sample had a mean of 11.4243 dollars and a standard deviation of 8.05794 dollars. Since the conditions for inference are satisfied, we can model the sampling distribution of the mean daily expense with a Student's $t$ model, with $14 - 1 = 13$ degrees of freedom, $t_{13}\left(10, \dfrac{8.05794}{\sqrt{14}}\right)$.

We will perform a one-sample $t$-test.

Since the *P*-value = 0.2600 is high, we fail to reject the null hypothesis. There is no evidence that the student's average spending is more than $10 per day.

$$t = \frac{\bar{y} - \mu_0}{SE(\bar{y})}$$

$$t = \frac{11.4243 - 10}{\frac{8.05794}{\sqrt{14}}}$$

$$t \approx 0.66$$

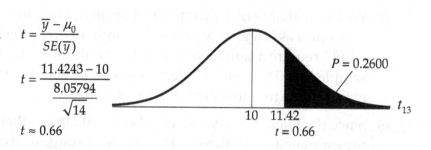

$P = 0.2600$

$t_{13}$

10   11.42

$t = 0.66$

### R4.47. Streams.

**Random condition:** The researchers randomly selected 172 streams.
**10% condition:** 172 is less than 10% of all streams.
**Success/Failure condition:** $n\hat{p} = 69$ and $n\hat{q} = 103$ are both greater than 10, so the sample is large enough.

Since the conditions are met, we can use a one-proportion *z*-interval to estimate the percentage of Adirondack streams with a shale substrate.

$$\hat{p} \pm z^* \sqrt{\frac{\hat{p}\hat{q}}{n}} = \left(\frac{69}{172}\right) \pm 1.960 \sqrt{\frac{\left(\frac{69}{172}\right)\left(\frac{103}{172}\right)}{172}} = (32.8\%, 47.4\%)$$

We are 95% confident that between 32.8% and 47.4% of Adirondack streams have a shale substrate.

### R4.49. Bread.

a) Since the histogram shows that the distribution of the number of loaves sold per day is skewed strongly to the right, we can't use the Normal model to estimate the number of loaves sold on the busiest 10% of days.

b) **Randomization condition:** Assume that these days are representative of all days.
**Nearly Normal condition:** The histogram is skewed strongly to the right. However, since the sample size is large, the Central Limit Theorem guarantees that the distribution of averages will be approximately Normal.

The days in the sample had a mean of 103 loaves sold and a standard deviation of 9 loaves sold. Since the conditions are satisfied, the sampling distribution of the mean can be modeled by a Student's *t*- model, with $100 - 1 = 99$ degrees of freedom. We will use a one-sample *t*-interval with 95% confidence for the mean number of loaves sold. (By hand, use $t_{50}^* \approx 2.403$ from the table.)

c) $$\bar{y} \pm t_{n-1}^* \left(\frac{s}{\sqrt{n}}\right) = 103 \pm t_{99}^* \left(\frac{9}{\sqrt{100}}\right) \approx (101.2,\ 104.8)$$

We are 95% confident that the mean number of loaves sold per day at the Clarksburg Bakery is between 101.2 and 104.8.

d) We know that in order to cut the margin of error in half, we need to a sample four times as large. If we allow a margin of error that is twice as wide, that would require a sample only one-fourth the size. In this case, our original sample is 100 loaves; so 25 loaves would be a sufficient number to estimate the mean with a margin of error twice as wide.

e) Since the interval is completely above 100 loaves, there is strong evidence that the estimate was incorrect. The evidence suggests that the mean number of loaves sold per day is greater than 100. This difference is statistically significant, but may not be practically significant. It seems like the owners made a pretty good estimate!

**R4.51. And it means?**

a) The margin of error is $\dfrac{(2391 - 1644)}{2} = \$373.50$.

b) The insurance agent is 95% confident that the mean loss claimed by clients after home burglaries is between $1644 and $2391.

c) 95% of all random samples of this size will produce intervals that contain the true mean loss claimed.

**R4.53. Recalls.**

Organize the information in a tree diagram.

a)

$$P(\text{recall}) = P(\text{American recall})$$
$$+ P(\text{Japanese recall})$$
$$+ P(\text{German recall})$$
$$= 0.014 + 0.002 + 0.001$$
$$= 0.017$$

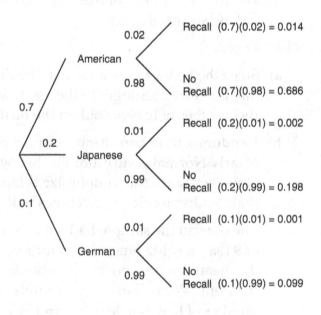

b)

$$P(\text{American} \mid \text{recall})$$
$$= \dfrac{P(\text{American and recall})}{P(\text{recall})}$$
$$= \dfrac{0.014}{0.014 + 0.002 + 0.001}$$

$$\approx 0.824$$

# Chapter 17 – Comparing Groups

## Section 17.1

**1. Canada.**

$$SE(\hat{p}_{Can} - \hat{p}_{Am}) = \sqrt{\frac{\hat{p}_{Can}\hat{q}_{Can}}{n_{Can}} + \frac{\hat{p}_{Am}\hat{q}_{Am}}{n_{Am}}} = \sqrt{\frac{\left(\frac{192}{960}\right)\left(\frac{768}{960}\right)}{960} + \frac{\left(\frac{170}{1250}\right)\left(\frac{1080}{1250}\right)}{1250}} = 0.0161$$

**3. Canada, deux.**

We are 95% confident that, based on these data, the proportion of foreign-born Canadians is between 3.24% and 9.56% more than the proportion of foreign-born Americans.

**5. Canada, trois.**

If we were to take repeated samples of these sizes of Canadians and Americans, and compute two-proportion confidence intervals, we would expect 95% of the intervals to contain the true difference in the proportions of foreign-born citizens.

## Section 17.2

**7. Canada, encore.**

We must assume the data were collected randomly and that the Americans selected are independent of the Canadians selected. Both assumptions should be met. All conditions for inference are met.

## Section 17.3

**9. Canada, test.**

a) $\hat{p}_{Can} - \hat{p}_{Am} = \dfrac{192}{960} - \dfrac{170}{1250} = 0.064$

b) $z = \dfrac{\hat{p}_{Can} - \hat{p}_{Am}}{\sqrt{\dfrac{\hat{p}_{Can}\hat{q}_{Can}}{n_{Can}} + \dfrac{\hat{p}_{Am}\hat{q}_{Am}}{n_{Am}}}} = \dfrac{\dfrac{192}{960} - \dfrac{170}{1250}}{\sqrt{\dfrac{\left(\frac{192}{960}\right)\left(\frac{768}{960}\right)}{960} + \dfrac{\left(\frac{170}{1250}\right)\left(\frac{1080}{1250}\right)}{1250}}} \approx 3.964$

(Using $SE_{pooled}(\hat{p}_{Can} - \hat{p}_{Am})$, $z \approx 4.03$)

c) Since the P-value is < 0.001, which is less than $\alpha = 0.05$, reject the null hypothesis. There is very strong evidence that the proportion of foreign born citizens is different in Canada than it is in the United States. According to this data, the proportion of foreign born Canadians is likely to be the higher of the two.

**Section 17.4**

**11. Cost of shopping.**

We must assume the samples were random or otherwise independent of each other. We also assume that the distributions are roughly normal, so it would be a good idea to check a histogram to make sure there isn't strong skewness or outliers.

**13. Cost of shopping, again.**

We are 95% confident that the mean purchase amount at Walmart is between $1.85 and $14.15 less than the mean purchase amount at Target.

**Section 17.5**

**15. Cost of shopping, once more.**

The difference is –$8 with an SE of 3.115, so the *t*-stat is –2.569. With 162.75 (or 163) df, the P-value is 0.011 which is less than 0.05. Reject the null hypothesis that the means are equal. There is evidence that the mean purchase amounts at the two stores are not the same. These data suggest that the mean purchase amount at Target is lower than the mean purchase price at Walmart.

**Section 17.7**

**17. Cost of shopping, yet again.**

The *t*-statistic is –2.561 using the pooled estimate of the standard deviation, 3.124. There are 163 df so the P-value is still 0.011. We reach the same conclusion as before. Because the sample standard deviations are nearly the same and the sample sizes are large, the pooled test is essentially the same as the two-sample *t*-test.

**19. Cost of shopping, once more.**

No. The two-sample test is almost always the safer choice. In this case, the variances are likely to be quite different. The purchase prices of Italian sports cars are much higher and may be more variable than the domestic prices. They should use the two-sample t-test.

**Chapter Exercises.**

**21. Online social networking.**

It is very unlikely that samples would show an observed difference this large if, in fact, there was no real difference between the proportion of American adults who visited Facebook on a daily basis in 2013 and the proportion of American adults who visited Facebook on a daily basis 2010.

## 23. Revealing information.

This test is not appropriate for these data, since the responses are not from independent groups, but are from the same individuals. The independent samples condition has been violated.

## 25. Gender gap.

a) This is a stratified random sample, stratified by gender.

b) We would expect the difference in proportions in the sample to be the same as the difference in proportions in the population, with the percentage of respondents with a favorable impression of the candidate 6 percentage points higher among males.

c) The standard deviation of the difference in proportions is:

$$SD(\hat{p}_M - \hat{p}_F) = \sqrt{\frac{\hat{p}_M \hat{q}_M}{n_M} + \frac{\hat{p}_F \hat{q}_F}{n_F}} = \sqrt{\frac{(0.59)(0.41)}{300} + \frac{(0.53)(0.47)}{300}} \approx 4\%$$

d)

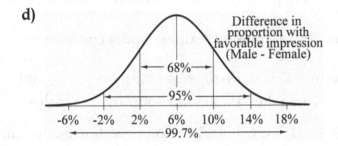

e) The campaign could certainly be misled by the poll. According to the model, a poll showing little difference could occur relatively frequently. That result is only 1.5 standard deviations below the expected difference in proportions.

## 27. Arthritis.

a) **Randomization condition:** Americans age 65 and older were selected randomly.
**Independent groups assumption:** The sample of men and the sample of women were drawn independently of each other.
**Success/Failure condition:** $n\hat{p}$ (men) = 411, $n\hat{q}$ (men) = 601, $n\hat{p}$ (women) = 535, and $n\hat{q}$ (women) = 527 are all greater than 10, so the samples are both large enough.

Since the conditions have been satisfied, we will find a two-proportion $z$-interval.

**b)**

$$\left(\hat{p}_F - \hat{p}_M\right) \pm z^* \sqrt{\frac{\hat{p}_F \hat{q}_F}{n_F} + \frac{\hat{p}_M \hat{q}_M}{n_M}} = \left(\tfrac{535}{1062} - \tfrac{411}{1012}\right) \pm 1.960 \sqrt{\frac{\left(\tfrac{535}{1062}\right)\left(\tfrac{527}{1062}\right)}{1062} + \frac{\left(\tfrac{411}{1012}\right)\left(\tfrac{601}{1012}\right)}{1012}}$$

$$= (0.055,\ 0.140)$$

**c)** We are 95% confident that the proportion of American women age 65 and older who suffer from arthritis is between 5.5% and 14.0% higher than the proportion of American men the same age who suffer from arthritis.

**d)** Since the interval for the difference in proportions of arthritis sufferers does not contain 0, there is strong evidence that arthritis is more likely to afflict women than men.

**29. Pets.**

**a)** $SE\left(\hat{p}_{Herb} - \hat{p}_{None}\right) = \sqrt{\dfrac{\hat{p}_{Herb} \hat{q}_{Herb}}{n_{Herb}} + \dfrac{\hat{p}_{None} \hat{q}_{None}}{n_{None}}} = \sqrt{\dfrac{\left(\tfrac{473}{827}\right)\left(\tfrac{354}{827}\right)}{827} + \dfrac{\left(\tfrac{19}{130}\right)\left(\tfrac{111}{130}\right)}{130}} = 0.035$

**b) Randomization condition:** Assume that the dogs studied were representative of all dogs.
**Independent groups assumption:** The samples were drawn independently of each other.
**Success/Failure condition:** $n\hat{p}$ (herb) = 473, $n\hat{q}$ (herb) = 354, $n\hat{p}$ (none) = 19, and $n\hat{q}$ (none) = 111 are all greater than 10, so the samples are both large enough.

Since the conditions have been satisfied, we will find a two-proportion z-interval.

$$\left(\hat{p}_{Herb} - \hat{p}_{None}\right) \pm z^* \sqrt{\frac{\hat{p}_{Herb} \hat{q}_{Herb}}{n_{Herb}} + \frac{\hat{p}_{None} \hat{q}_{None}}{n_{None}}}$$

$$= \left(\tfrac{473}{827} - \tfrac{19}{130}\right) \pm 1.960 \sqrt{\frac{\left(\tfrac{473}{827}\right)\left(\tfrac{354}{827}\right)}{827} + \frac{\left(\tfrac{19}{130}\right)\left(\tfrac{111}{130}\right)}{130}} = (0.356,\ 0.495)$$

**c)** We are 95% confident that the proportion of pets with a malignant lymphoma in homes where herbicides are used is between 35.6 and 49.5 percentage points higher than the proportion with lymphoma in homes where no pesticides are used.

**31. Prostate cancer.**

**a)** This is an experiment. Men were randomly assigned to have surgery or not.

b) **Randomization condition:** Men were randomly assigned to treatment groups.
**Independent groups assumption:** The survival rates of the two groups are not related.
**Success/Failure condition:** $n\hat{p}$ ( no surgery) = 31 , $n\hat{q}$ ( no surgery) = 317, $n\hat{p}$ (surgery) = 16, and $n\hat{q}$ (surgery) = 331 are all greater than 10, so the samples are both large enough.

Since the conditions have been satisfied, we will find a two-proportion $z$-interval.

$$\left(\hat{p}_{None} - \hat{p}_{Surg}\right) \pm z^* \sqrt{\frac{\hat{p}_{None}\hat{q}_{None}}{n_{None}} + \frac{\hat{p}_{Surg}\hat{q}_{Surg}}{n_{Surg}}}$$

$$= \left(\frac{31}{348} - \frac{16}{347}\right) \pm 1.960 \sqrt{\frac{\left(\frac{31}{348}\right)\left(\frac{317}{348}\right)}{348} + \frac{\left(\frac{16}{347}\right)\left(\frac{331}{347}\right)}{347}} = (0.006, 0.080)$$

c) We are 95% confident that the survival rate of patients who have surgery for prostate cancer is between 0.6 and 8.0 percentage points higher than the survival rate of patients who do not have surgery. Since the interval is completely above zero, there is evidence that surgery may be effective in preventing death from prostate cancer.

**33. Ear infections.**

a) **Randomization condition:** The babies were randomly assigned to the two treatment groups.
**Independent groups assumption:** The groups were assigned randomly, so the groups are not related.
**Success/Failure condition:** $n\hat{p}$ (vaccine) = 333, $n\hat{q}$ (vaccine) = 2122, $n\hat{p}$ (none) = 499, and $n\hat{q}$ (none) = 1953 are all greater than 10, so the samples are both large enough.

Since the conditions have been satisfied, we will find a two-proportion $z$-interval.

b) $\left(\hat{p}_{None} - \hat{p}_{Vacc}\right) \pm z^* \sqrt{\dfrac{\hat{p}_{None}\hat{q}_{None}}{n_{None}} + \dfrac{\hat{p}_{Vacc}\hat{q}_{Vacc}}{n_{Vacc}}}$

$$= \left(\frac{499}{2452} - \frac{333}{2455}\right) \pm 1.960 \sqrt{\frac{\left(\frac{499}{2452}\right)\left(\frac{1953}{2452}\right)}{2452} + \frac{\left(\frac{333}{2455}\right)\left(\frac{2122}{2455}\right)}{2455}} = (0.047, 0.089)$$

c) We are 95% confident that the proportion of unvaccinated babies who develop ear infections is between 4.7 and 8.9 percentage points higher than the proportion of vaccinated babies who develop ear infections. The vaccinations appear to be effective, especially considering the 20% infection rate among the unvaccinated. A reduction of 5% to 9% is meaningful.

### 35. Another ear infection.

**a)**  $H_0$ : The proportion of vaccinated babies who get ear infections is the same as the proportion of unvaccinated babies who get ear infections.

$$\left(p_{Vacc} = p_{None} \quad \text{or} \quad p_{Vacc} - p_{None} = 0\right)$$

$H_A$ : The proportion of vaccinated babies who get ear infections is the lower than the proportion of unvaccinated babies who get ear infections.

$$\left(p_{Vacc} < p_{None} \quad \text{or} \quad p_{Vacc} - p_{None} < 0\right)$$

**b)**  Since 0 is not in the confidence interval, reject the null hypothesis. There is evidence that the vaccine reduces the rate of ear infections.

**c)**  If we think that the vaccine really reduces the rate of ear infections and it really does not reduce the rate of ear infections, then we have committed a Type I error.

**d)**  Babies would be given ineffective vaccines.

### 37. Teen smoking.

**a)**  This is a prospective observational study.

**b)**  $H_0$ : The proportion of teen smokers among the group whose parents disapprove of smoking is the same as the proportion of teen smokers among the group whose parents are lenient about smoking.

$$\left(p_{Dis} = p_{Len} \quad \text{or} \quad p_{Dis} - p_{Len} = 0\right)$$

$H_A$ : The proportion of teen smokers among the group whose parents disapprove of smoking is different than the proportion of teen smokers among the group whose parents are lenient about smoking.

$$\left(p_{Dis} \neq p_{Len} \quad \text{or} \quad p_{Dis} - p_{Len} \neq 0\right)$$

**c)**  **Randomization condition:** Assume that the teens surveyed are representative of all teens.
**Independent groups assumption:** The groups were surveyed independently.
**Success/Failure condition:** $n\hat{p}$ (disapprove) = 54, $n\hat{q}$ (disapprove) = 230, $n\hat{p}$ (lenient) = 11, and $n\hat{q}$ (lenient) = 30 are all greater than 10, so the samples are both large enough.

Since the conditions have been satisfied, we will model the sampling distribution of the difference in proportion with a Normal model with mean 0 and standard deviation estimated by

$$SE\left(\hat{p}_{Dis} - \hat{p}_{Len}\right) = \sqrt{\frac{\hat{p}_{Dis}\hat{q}_{Dis}}{n_{Dis}} + \frac{\hat{p}_{Len}\hat{q}_{Len}}{n_{Len}}} = \sqrt{\frac{\left(\frac{54}{284}\right)\left(\frac{230}{284}\right)}{284} + \frac{\left(\frac{11}{41}\right)\left(\frac{30}{41}\right)}{41}} = 0.0730.$$

**d)** The observed difference between the proportions is $0.190 - 0.268 = -0.078$.

Since the P-value = 0.28 is high, we fail to reject the null hypothesis. There is little evidence to suggest that parental attitudes influence teens' decisions to smoke.

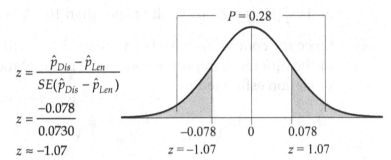

$$z = \frac{\hat{p}_{Dis} - \hat{p}_{Len}}{SE(\hat{p}_{Dis} - \hat{p}_{Len})}$$

$$z = \frac{-0.078}{0.0730}$$

$$z \approx -1.07$$

**e)** If there is no difference in the proportions, there is about a 28% chance of seeing the observed difference or larger by natural sampling variation.

**f)** If teens' decisions about smoking *are* influenced, we have committed a Type II error.

**g)** Since the conditions have already been satisfied in a previous exercise, we will find a two-proportion z-interval.

$$\left(\hat{p}_{Dis} - \hat{p}_{Len}\right) \pm z^* \sqrt{\frac{\hat{p}_{Dis}\hat{q}_{Dis}}{n_{Dis}} + \frac{\hat{p}_{Len}\hat{q}_{Len}}{n_{Len}}}$$

$$= \left(\tfrac{54}{284} - \tfrac{11}{41}\right) \pm 1.960 \sqrt{\frac{\left(\tfrac{54}{284}\right)\left(\tfrac{230}{284}\right)}{284} + \frac{\left(\tfrac{11}{41}\right)\left(\tfrac{30}{41}\right)}{41}} = \left(-0.221, 0.065\right)$$

**h)** We are 95% confident that the proportion of teens whose parents disapprove of smoking who will eventually smoke is between 22.1 percentage points lower and 6.5 percentage points higher than for teens with parents who are lenient about smoking.

**i)** We expect 95% of random samples of this size to produce intervals that contain the true difference between the proportions.

## 39. Birthweight.

**a)** $H_0$: The proportion of low birthweight is the same. $\left(p_{Exp} = p_{Not} \text{ or } p_{Exp} - p_{Not} = 0\right)$

$H_A$: The proportion of low birthweight is higher for women exposed to soot and ash. $\left(p_{Exp} > p_{Not} \text{ or } p_{Exp} - p_{Not} > 0\right)$

**Randomization condition:** Assume the women are representative of all women.
**Independent groups assumption:** The groups don't appear to be associated, with respect to soot and ash exposure, but all of the women were in New York. There may be a confounding variable explaining any relationship between exposure and birthweight.

**Success/Failure condition:** $n\hat{p}$ (Exposed) = 15, $n\hat{q}$ (Exposed) = 167, $n\hat{p}$ (Not) = 92, and $n\hat{q}$ (Not) = 2208 are all greater than 10. All of the samples are large enough.

Since the conditions have been satisfied, we will model the sampling distribution of the difference in proportion with a Normal model with mean 0 and standard deviation estimated by

$$SE\left(\hat{p}_{Exp} - \hat{p}_{Not}\right) = \sqrt{\frac{\hat{p}_{Exp}\hat{q}_{Exp}}{n_{Exp}} + \frac{\hat{p}_{Not}\hat{q}_{Not}}{n_{Not}}} = \sqrt{\frac{\left(\frac{15}{182}\right)\left(\frac{167}{182}\right)}{182} + \frac{\left(\frac{92}{2300}\right)\left(\frac{2208}{2300}\right)}{2300}} \approx 0.020790.$$

The observed difference between the proportions is: $\left(\frac{15}{182}\right) - \left(\frac{92}{2300}\right) = 0.04242$

Since the P-value = 0.021 is low, we reject the null hypothesis. There is strong evidence that the proportion of low birthweight babies is higher in the women exposed to soot and ash after the World Trade Center attacks.

$$z = \frac{\hat{p}_{Exp} - \hat{p}_{Not}}{SE(\hat{p}_{Exp} - \hat{p}_{Not})}$$

$$z = \frac{0.04242}{0.020790}$$

$$z \approx 2.04$$

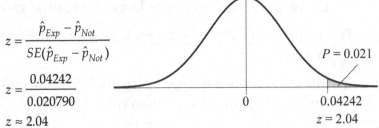

$P = 0.021$

$0$      $0.04242$

$z = 2.04$

**b)** $\left(\hat{p}_{Exp} - \hat{p}_{Not}\right) \pm z^* \sqrt{\dfrac{\hat{p}_{Exp}\hat{q}_{Exp}}{n_{Exp}} + \dfrac{\hat{p}_{Not}\hat{q}_{Not}}{n_{Not}}}$

$$= \left(\frac{15}{182} - \frac{92}{2300}\right) \pm 1.960 \sqrt{\frac{\left(\frac{15}{182}\right)\left(\frac{167}{182}\right)}{182} + \frac{\left(\frac{92}{2300}\right)\left(\frac{2208}{2300}\right)}{2300}} = (0.002, 0.083)$$

We are 95% confident that the proportion of low birthweight babies is between 0.2 and 8.3 percentage points higher for mothers exposed to soot and ash after the World Trade Center attacks, than the proportion of low birthweight babies for mothers not exposed.

**41. Mammograms.**

**a)** $H_0$: The proportion of deaths from breast cancer is the same for women who never had a mammogram as for women who had mammograms.
$\left(p_N = p_M \text{ or } p_N - p_M = 0\right)$

$H_A$: The proportion of deaths from breast cancer is greater for women who never had a mammogram than for women who had mammograms.
$\left(p_N > p_M \text{ or } p_N - p_M > 0\right)$

**Randomization condition:** Assume the women are representative of all women. **Independent groups assumption:** We must assume that these groups are independent.

**Success/Failure condition:** $n\hat{p}$ (never) = 196, $n\hat{q}$ (never) = 30,369, $n\hat{p}$ (mammogram) = 153, and $n\hat{q}$ (mammogram) = 29,978 are all greater than 10, so both samples are large enough.

Since the conditions have been satisfied, we will model the sampling distribution of the difference in proportion with a Normal model with mean 0 and standard deviation estimated by:

$$SE_{pooled}\left(\hat{p}_N - \hat{p}_M\right) = \sqrt{\frac{\hat{p}_{pooled}\hat{q}_{pooled}}{n_N} + \frac{\hat{p}_{pooled}\hat{q}_{pooled}}{n_M}}$$

$$= \sqrt{\frac{\left(\frac{349}{60,696}\right)\left(\frac{60,347}{60,696}\right)}{30,565} + \frac{\left(\frac{349}{60,696}\right)\left(\frac{60,347}{60,696}\right)}{30,131}} \approx 0.000614$$

We could use the unpooled standard error. There is no practical difference.

The observed difference between the proportions is:
0.006413 – 0.005078 = 0.001335.

Since the *P*-value = 0.0148 is low, we reject the null hypothesis. There is strong evidence that the breast cancer mortality rate is higher among women that have never

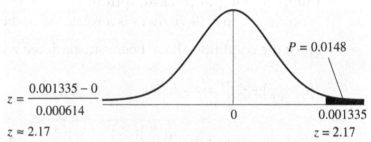

$z = \dfrac{0.001335 - 0}{0.000614}$

$z \approx 2.17$

had a mammogram. However, the large sample sizes involved may have yielded a result that is statistically significant, but not practically significant. These data suggest a difference in mortality rate of only about 0.1 percentage points.

**b)** If there is actually no difference in the mortality rates, we have committed a Type I error.

## 43. Pain.

**a)** **Randomization condition:** The patients were randomly selected AND randomly assigned to treatment groups. If that's not random enough for you, I don't know what is!

**Success/Failure condition:** $n\hat{p}$ (A) = 84, $n\hat{q}$ (A) = 28, $n\hat{p}$ (B) = 66, and $n\hat{q}$ (B) = 42 are all greater than 10, so both samples are large enough.

Since the conditions are met, we can use a one-proportion *z*-interval to estimate the percentage of patients who may get relief from medication A.

$$\hat{p} \pm z^* \sqrt{\frac{\hat{p}\hat{q}}{n}} = \left(\frac{84}{112}\right) \pm 1.960\sqrt{\frac{\left(\frac{84}{112}\right)\left(\frac{28}{112}\right)}{112}} = (67.0\%, 83.0\%)$$

We are 95% confident that between 67.0% and 83.0% of patients with joint pain will find medication A to be effective.

**b)** Since the conditions were met in part a, we can use a one-proportion z-interval to estimate the percentage of patients who may get relief from medication B.

$$\hat{p} \pm z^* \sqrt{\frac{\hat{p}\hat{q}}{n}} = \left(\frac{66}{108}\right) \pm 1.960 \sqrt{\frac{\left(\frac{66}{108}\right)\left(\frac{42}{108}\right)}{108}} = (51.9\%, 70.3\%)$$

We are 95% confident that between 51.9% and 70.3% of patients with joint pain will find medication B to be effective.

**c)** The 95% confidence intervals overlap, which might lead one to believe that there is no evidence of a difference in the proportions of people who find each medication effective.  However, if one was lead to believe that, one should proceed to part…

**d)** Most of the conditions were checked in part a.  We only have one more to check: **Independent groups assumption:** The groups were assigned randomly, so there is no reason to believe there is a relationship between them.

Since the conditions have been satisfied, we will find a two-proportion z-interval.

$$\left(\hat{p}_A - \hat{p}_B\right) \pm z^* \sqrt{\frac{\hat{p}_A\hat{q}_A}{n_A} + \frac{\hat{p}_B\hat{q}_B}{n_B}}$$

$$= \left(\frac{84}{112} - \frac{66}{108}\right) \pm 1.960 \sqrt{\frac{\left(\frac{84}{112}\right)\left(\frac{28}{112}\right)}{112} + \frac{\left(\frac{66}{108}\right)\left(\frac{42}{108}\right)}{112}} = (0.017, 0.261)$$

We are 95% confident that the proportion of patients with joint pain who will find medication A effective is between 1.70 and 26.1 percentage points higher than the proportion of patients who will find medication B effective.

**e)** The interval does not contain zero.  There is evidence that medication A is more effective than medication B.

**f)** The two-proportion method is the proper method.  By attempting to use two, separate, confidence intervals, you are adding standard deviations when looking for a difference in proportions.  We know from our previous studies that *variances* add when finding the standard deviation of a difference.  The two-proportion method does this.

**45. Convention bounce.**

**Randomization condition:** The polls were conducted randomly.
**Independent groups assumption:** The groups were chosen independently.
**Success/Failure condition:** $n\hat{p}$ (after) = 705, $n\hat{q}$ (after) = 795, $n\hat{p}$ (before) = 735, and $n\hat{q}$ (women) = 765 are all greater than 10, so both samples are large enough.

Since the conditions have been satisfied, we will find a two-proportion z-interval.

$$\left(\hat{p}_{After} - \hat{p}_{Before}\right) \pm z^* \sqrt{\frac{\hat{p}_{After}\hat{q}_{After}}{n_{After}} + \frac{\hat{p}_{Before}\hat{q}_{Before}}{n_{Before}}}$$

$$= (0.49 - 0.47) \pm 1.960 \sqrt{\frac{(0.49)(0.51)}{1500} + \frac{(0.47)(0.53)}{1500}} = (-0.016, 0.056)$$

We can be 95% confident that the proportion of likely voters who favored John Kerry was between 1.6 percentage points lower and 5.6 percentage points higher after the convention, compared to before the convention. Since zero is contained in the interval, it is plausible that there was no difference in Kerry support. The poll showed no evidence of a convention bounce.

### 47. Sensitive men.

$H_0$: The proportion of 18-24-year-old men who are comfortable talking about their problems is the same as the proportion of 25-34-year old men.
$$\left(p_{Young} = p_{Old} \text{ or } p_{Young} - p_{Old} = 0\right)$$

$H_A$: The proportion of 18-24-year-old men who are comfortable talking about their problems is higher than the proportion of 25-34-year old men.
$$\left(p_{Young} > p_{Old} \text{ or } p_{Young} - p_{Old} > 0\right)$$

**Randomization condition:** We assume the respondents were chosen randomly.
**Independent groups assumption:** The groups were chosen independently.
**Success/Failure condition:** $n\hat{p}$ (young) = 80, $n\hat{q}$ (young) = 49, $n\hat{p}$ (old) = 98, and $n\hat{q}$ (old) = 86 are all greater than 10, so both samples are large enough.

Since the conditions have been satisfied, we will model the sampling distribution of the difference in proportion with a Normal model with mean 0 and standard deviation using the pooled proportion estimated by:

$$SE_{pooled}\left(\hat{p}_{Young} - \hat{p}_{Old}\right) = \sqrt{\frac{\hat{p}_{pooled}\hat{q}_{pooled}}{n_Y} + \frac{\hat{p}_{pooled}\hat{q}_{pooled}}{n_O}} = \sqrt{\frac{\left(\frac{178}{313}\right)\left(\frac{135}{313}\right)}{129} + \frac{\left(\frac{178}{313}\right)\left(\frac{135}{313}\right)}{184}} \approx 0.05687$$

We could use the unpooled standard error. There is no practical difference.

The observed difference between the proportions is: 0.620 – 0.533 = 0.087.

Since the P-value = 0.0619 is high, we fail to reject the null hypothesis. There is little evidence that the proportion of 18-24-year-old men who are comfortable talking about their problems is higher than the proportion of 25-34-year-old men who are comfortable. *Time* magazine's interpretation is questionable.

$$z = \frac{0.087 - 0}{0.05687}$$

$$z \approx 1.54$$

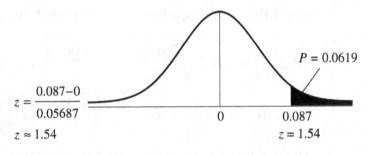

### 49. Food preference.

$H_0$: The proportion of people who agree with the statement is the same in rural and urban areas. $\left( p_{Urban} = p_{Rural} \text{ or } p_{Urban} - p_{Rural} = 0 \right)$

$H_A$: The proportion of people who agree with the statement differs in rural and urban areas. $\left( p_{Urban} \neq p_{Rural} \text{ or } p_{Urban} - p_{Rural} \neq 0 \right)$

**Randomization condition:** The respondents were chosen randomly.
**Independent groups assumption:** The groups were chosen independently.
**Success/Failure condition:** $n\hat{p}$ (urban) = 417, $n\hat{q}$ (urban) = 229, $n\hat{p}$ (rural) = 78, and $n\hat{q}$ (rural) = 76 are all greater than 10, so both samples are large enough.

Since the conditions have been satisfied, we will model the sampling distribution of the difference in proportion with a Normal model with mean 0 and standard deviation using the pooled proportion estimated by:

$$SE_{pooled}\left(\hat{p}_{Urban} - \hat{p}_{Rural}\right) = \sqrt{\frac{\hat{p}_{pooled}\hat{q}_{pooled}}{n_{Urban}} + \frac{\hat{p}_{pooled}\hat{q}_{pooled}}{n_{Rural}}} = \sqrt{\frac{\left(\frac{495}{800}\right)\left(\frac{305}{800}\right)}{646} + \frac{\left(\frac{495}{800}\right)\left(\frac{305}{800}\right)}{154}} \approx 0.0436$$

We could use the unpooled standard error. There is no practical difference.

The observed difference between the proportions is: $\frac{417}{646} - \frac{78}{154} \approx 0.139$

Since the P-value = 0.0014 is low, reject the null hypothesis. There is evidence that the proportion of people who agree with the statement is not the same in urban and rural areas. These data suggest that the proportion is higher in urban areas than in rural areas.

$$z = \frac{0.139 - 0}{0.0436}$$

$$z \approx 3.19$$

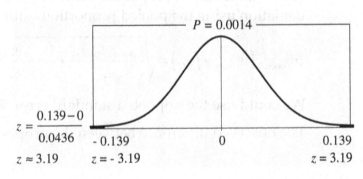

### 51. Hot dogs.

Yes, the 95% confidence interval would contain 0. The high P-value means that we lack evidence of a difference, so 0 is a possible value for $\mu_{Meat} - \mu_{Beef}$.

### 53. Hot dogs, second helping.

a) Plausible values for $\mu_{Meat} - \mu_{Beef}$ are all negative, so the mean fat content is probably higher for beef hot dogs.

b) The fact that the confidence interval does not contain 0 indicates that the difference is significant.

c) The corresponding alpha level is 10%.

### 55. Hot dogs, last one.

a) False. The confidence interval is about means, not about individual hot dogs.

b) False. The confidence interval is about means, not about individual hot dogs.

c) True.

d) False. Confidence intervals based on other samples will also try to estimate the true difference in population means. There's not reason to expect other samples to conform to this result.

e) True.

### 57. Learning math.

a) The margin of error of this confidence interval is (11.427 – 5.573)/2 = 2.927 points.

b) The margin of error for a 98% confidence interval would have been larger. The critical value of $t^*$ is larger for higher confidence levels. We need a wider interval to increase the likelihood that we catch the true mean difference in test scores within our interval. In other words, greater confidence comes at the expense of precision.

c) We are 95% confident that the mean score for the CPMP math students will be between 5.573 and 11.427 points higher on this assessment than the mean score of the traditional students.

d) Since the entire interval is above 0, there is strong evidence that students who learn with CPMP will have higher mean scores is algebra than those in traditional programs.

## 59. CPMP, again.

a) $H_0$: The mean score of CPMP students is the same as the mean score of traditional students. $\left(\mu_C = \mu_T \text{ or } \mu_C - \mu_T = 0\right)$

$H_A$: The mean score of CPMP students is different from the mean score of traditional students. $\left(\mu_C \neq \mu_T \text{ or } \mu_C - \mu_T \neq 0\right)$

b) **Independent groups assumption:** Scores of students from different classes should be independent.
**Randomization condition:** Although not specifically stated, classes in this experiment were probably randomly assigned to either CPMP or traditional curricula.
**Nearly Normal condition:** We don't have the actual data, so we can't check the distribution of the sample. However, the samples are large. The Central Limit Theorem allows us to proceed.

Since the conditions are satisfied, we can use a two-sample $t$-test with 583 degrees of freedom (from the computer).

c) If the mean scores for the CPMP and traditional students are really equal, there is less than a 1 in 10,000 chance of seeing a difference as large or larger than the observed difference just from natural sampling variation.

d) Since the $P$-value < 0.0001, reject the null hypothesis. There is strong evidence that the CPMP students have a different mean score than the traditional students. The evidence suggests that the CPMP students have a higher mean score.

## 61. Commuting.

a) **Independent groups assumption:** Since the choice of route was determined at random, the commuting times for Route A are independent of the commuting times for Route B.
**Randomization condition:** The man randomly determined which route he would travel on each day.
**Nearly Normal condition:** The histograms of travel times for the routes are roughly unimodal and symmetric. (Given)

Since the conditions are satisfied, it is appropriate to model the sampling distribution of the difference in means with a Student's $t$-model, with 33.1 degrees of freedom (from the approximation formula). We will construct a two-sample $t$-interval, with 95% confidence.

$$\left(\bar{y}_B - \bar{y}_A\right) \pm t_{df}^* \sqrt{\frac{s_B^2}{n_B} + \frac{s_A^2}{n_A}} = \left(43 - 40\right) \pm t_{33.1}^* \sqrt{\frac{2^2}{20} + \frac{3^2}{20}} \approx \left(1.36, \ 4.64\right)$$

We are 95% confident that Route B has a mean commuting time between 1.36 and 4.64 minutes longer than the mean commuting time of Route A.

**b)** Since 5 minutes is beyond the high end of the interval, there is no evidence that the Route B is an average of 5 minutes longer than Route A. It appears that the old-timer may be exaggerating the average difference in commuting time.

## 63. View of the water.

**Independent groups assumption:** Since the 170 properties were randomly selected, the groups should be independent.
**Randomization condition:** The 170 properties were selected randomly.
**Nearly Normal condition:** The boxplots of sale prices are roughly symmetric. The plots show several outliers, but the sample sizes are large. The Central Limit Theorem allows us to proceed.

Since the conditions are satisfied, it is appropriate to model the sampling distribution of the difference in means with a Student's *t*-model, with 105.48 degrees of freedom (from the approximation formula). We will construct a two-sample *t*-interval, with 95% confidence.

$$(\bar{y}_W - \bar{y}_N) \pm t^*_{df} \sqrt{\frac{s^2_W}{n_W} + \frac{s^2_N}{n_N}}$$

$$= (319,906.40 - 219,896.60) \pm t^*_{105.48} \sqrt{\frac{153,303.80^2}{70} + \frac{94,627.15^2}{100}}$$

$$\approx (\$59121, \$140898)$$

We are 95% confident that waterfront property has a mean selling price that is between \$59,121 and \$140,898 higher, on average, than non-waterfront property.

## 65. Cereal sugar.

**Independent groups assumption:** The percentage of sugar in the children's cereals is unrelated to the percentage of sugar in adult's cereals.
**Randomization condition:** It is reasonable to assume that the cereals are representative of all children's cereals and adult cereals, in regard to sugar content.
**Nearly Normal condition:** The histogram of adult cereal sugar content is skewed to the right, but the sample sizes are of reasonable size. The Central Limit Theorem allows us to proceed.

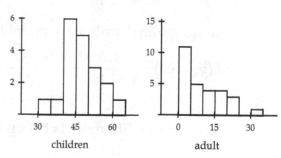

Since the conditions are satisfied, it is appropriate to model the sampling distribution of the difference in means with a Student's *t*-model, with 42 degrees of freedom (from the approximation formula). We will construct a two-sample *t*-interval, with 95% confidence.

$$(\bar{y}_C - \bar{y}_A) \pm t^*_{df}\sqrt{\frac{s_C^2}{n_C} + \frac{s_A^2}{n_A}} = (46.8 - 10.1536) \pm t^*_{42}\sqrt{\frac{6.41838^2}{19} + \frac{7.61239^2}{28}} \approx (32.49, 40.80)$$

We are 95% confident that children's cereals have a mean sugar content that is between 32.49% and 40.80% higher than the mean sugar content of adult cereals.

**67. Reading.**

$H_0$: The mean reading comprehension score of students who learn by the new method is the same as the mean score of students who learn by traditional methods. $\left(\mu_N = \mu_T \text{ or } \mu_N - \mu_T = 0\right)$

$H_A$: The mean reading comprehension score of students who learn by the new method is greater than the mean score of students who learn by traditional methods. $\left(\mu_N > \mu_T \text{ or } \mu_N - \mu_T > 0\right)$

**Independent groups assumption:** Student scores in one group should not have an impact on the scores of students in the other group.
**Randomization condition:** Students were randomly assigned to classes.
**Nearly Normal condition:** The stem-and-leaf plots show distributions of scores that are unimodal and symmetric.

Since the conditions are satisfied, it is appropriate to model the sampling distribution of the difference in means with a Student's *t*-model, with 37.3 degrees of freedom (from the approximation formula). We will perform a two-sample *t*-test. We know:

$$\bar{y}_N = 51.7222 \qquad \bar{y}_T = 41.8182$$
$$s_N = 11.7062 \qquad s_T = 16.5979$$
$$n_N = 18 \qquad n_T = 22$$

The sampling distribution model has mean 0, with standard error:

$$SE(\bar{y}_N - \bar{y}_T) = \sqrt{\frac{11.7062^2}{18} + \frac{16.5979^2}{22}} \approx 4.487.$$

The observed difference between the mean scores is $51.7222 - 41.8182 \approx 9.904$.

Since the *P*-value = 0.0168 is low, we reject the null hypothesis. There is evidence that the students taught using the new

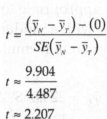

$$t = \frac{(\bar{y}_N - \bar{y}_T) - (0)}{SE(\bar{y}_N - \bar{y}_T)}$$

$$t \approx \frac{9.904}{4.487}$$

$$t \approx 2.207$$

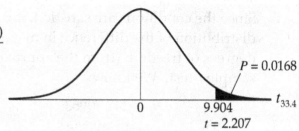

activities have a higher mean score on the reading comprehension test than the students taught using traditional methods.

**69. Cholesterol and gender.**

a) Boxplots are at the right.

b) Since there is a great deal of overlap between the cholesterol levels for men and women, it does not appear that one gender has higher average levels of cholesterol.

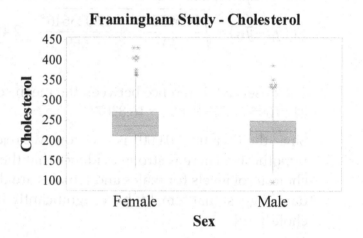

c) $H_0$: The mean cholesterol level is the same for females and males.
$$\left( \mu_F = \mu_M \ \text{ or } \ \mu_F - \mu_M = 0 \right)$$

$H_A$ The mean cholesterol level is different for females and males.
$$\left( \mu_F \neq \mu_M \ \text{ or } \ \mu_F - \mu_M \neq 0 \right)$$

**Independent groups assumption:** Female cholesterol levels should not affect male cholesterol
**Randomization condition:** Participants in the Framingham survey should be representative of males and females in general.
**Nearly Normal condition:** The distribution of cholesterol levels are both generally unimodal and symmetric, though each has a large number of outliers.

Since the conditions are satisfied, it is appropriate to model the sampling distribution of the difference in means with a Student's $t$-model, with 1401 degrees of freedom (from the approximation formula). We will perform a two-sample $t$-test.  We know:

$$\bar{y}_F = 242.70963 \qquad \bar{y}_M = 225.87892$$
$$s_F = 48.505208 \qquad s_M = 42.023646$$
$$n_F = 737 \qquad n_M = 669$$

The sampling distribution model has mean 0, with standard error:

$$SE(\bar{y}_F - \bar{y}_M) = \sqrt{\frac{48.505208^2}{737} + \frac{42.023646^2}{669}} \approx 2.4149704.$$

The observed difference between the mean scores is $242.70963 - 225.87892 \approx 16.83071$.

Since the $P$-value $< 0.0001$ is very low, we reject the null hypothesis.  There is strong evidence that the mean cholesterol levels for males and females are different. These data suggest that females have significantly higher mean cholesterol.

$$t = \frac{(\bar{y}_F - \bar{y}_M) - (0)}{SE(\bar{y}_F - \bar{y}_M)}$$
$$t \approx \frac{16.83071}{2.4149704}$$
$$t \approx 6.97$$

$$(\bar{y}_F - \bar{y}_M) \pm t_{df}^* \sqrt{\frac{s_F^2}{n_F} + \frac{s_M^2}{n_M}}$$

$$= (242.70963 - 225.87892) \pm t_{1401}^* \sqrt{\frac{48.505208^2}{737} + \frac{42.023646^2}{669}} \approx (12.1, \ 21.6)$$

We are 95% confident that the interval 12.1 to 21.6 contains the true mean difference in the cholesterol levels of females and males. In other words, we are 95% confident that females have a mean cholesterol level between 12.1 and 21.6 mg/dL higher than the mean cholesterol level of men.

**d)** Our conclusion is different than our conclusion in part b. The evidence of a difference in mean cholesterol level is much stronger than it appeared in the boxplot. The large sample sizes make us very confident in our predictions of the means.

**e)** Removing the outliers would not change our conclusion. Without the outliers, the standard error of the means would be smaller, since the cholesterol levels would be less variable, and the means would not change much. The value of the test statistic would not change that much.

## 71. Home runs 2016.

**a)** The boxplots of the average number of home runs hit at the ballparks in the two leagues are at the right. Both distributions appear at least roughly symmetric, with no outliers.

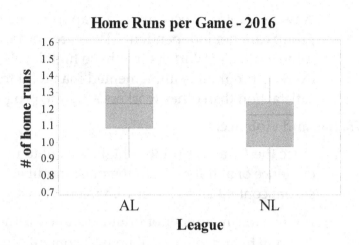

**b)** $\bar{y} \pm t^*_{n-1}\left(\dfrac{s}{\sqrt{n}}\right) = 1.21624 \pm t^*_{14}\left(\dfrac{0.17736}{\sqrt{15}}\right) \approx (1.12, 1.31)$

We are 95% confident that the mean number of home runs hit per game in American League stadiums is between 1.12 and 1.31.

**c)** The average of 1.26 home runs hit per game in Coors Field is not unusual. It isn't an outlier on the boxplot, and isn't even the highest average in the National League.

**d)** If you attempt to use two confidence intervals to assess a difference in means, you are actually adding standard deviations. But it's the variances that add, not the standard deviations. The two-sample difference of means procedure takes this into account.

**e)**

$(\bar{y}_A - \bar{y}_N) \pm t^*_{df}\sqrt{\dfrac{s^2_A}{n_A} + \dfrac{s^2_N}{n_N}}$

$= (1.21624 - 1.09495) \pm t^*_{27.55}\sqrt{\dfrac{0.17736^2}{15} + \dfrac{0.20179^2}{15}} \approx (-0.021, 0.263)$

**f)** We are 95% confident that the mean number of home runs in American League stadiums is between 0.021 home runs lower and 0.263 home runs higher than the mean number of home runs in National League stadiums.

**g)** Since the interval contains 0, there is no evidence of a difference in the mean number of home runs hit per game in the stadiums of the two leagues.

### 73. Job satisfaction.

A two-sample *t*-procedure is not appropriate for these data, because the two groups are not independent. They are before and after satisfaction scores for the same workers. Workers that have high levels of job satisfaction before the exercise program is implemented may tend to have higher levels of job satisfaction than other workers after the program as well.

### 75. Sex and violence.

a) Since the *P*-value = 0.136 is high, we fail to reject the null hypothesis. There is no evidence of a difference in the mean number of brands recalled by viewers of sexual content and viewers of violent content.

b) H₀: The mean number of brands recalled is the same for viewers of sexual content and viewers of neutral content. $\left( \mu_S = \mu_N \text{ or } \mu_S - \mu_N = 0 \right)$

   Hₐ: The mean number of brands recalled is different for viewers of sexual content and viewers of neutral content. $\left( \mu_S \ne \mu_N \text{ or } \mu_S - \mu_N \ne 0 \right)$

   **Independent groups assumption:** Recall of one group should not affect recall of another.
   **Randomization condition:** Subjects were randomly assigned to groups.
   **Nearly Normal condition:** The samples are large.

   Since the conditions are satisfied, it is appropriate to model the sampling distribution of the difference in means with a Student's *t*-model, with 214 degrees of freedom (from the approximation formula). We will perform a two-sample *t*-test.

   The sampling distribution model has mean 0, with standard error:

   $$SE(\bar{y}_S - \bar{y}_N) = \sqrt{\frac{1.76^2}{108} + \frac{1.77^2}{108}} \approx 0.24 \, .$$

   The observed difference between the mean scores is 1.71 – 3.17 = – 1.46.

   Since the *P*-value = $5.5 \times 10^{-9}$ is low, we reject the null hypothesis. There is strong evidence that the mean number of brand names recalled is different for viewers of sexual content and viewers of neutral content. The evidence suggests that viewers of neutral ads remember more brand names on average than viewers of sexual content.

   $$t = \frac{(\bar{y}_S - \bar{y}_N) - (0)}{SE(\bar{y}_S - \bar{y}_N)}$$

   $$t \approx \frac{-1.46}{0.24}$$

   $$t \approx -6.08$$

## 77. Hungry?

$H_0$: The mean number of ounces of ice cream people scoop is the same for large and small bowls. $\left(\mu_{big} = \mu_{small} \quad \text{or} \quad \mu_{big} - \mu_{small} = 0\right)$

$H_A$: The mean number of ounces of ice cream people scoop is the different for large and small bowls. $\left(\mu_{big} \neq \mu_{small} \quad \text{or} \quad \mu_{big} - \mu_{small} \neq 0\right)$

**Independent groups assumption:** The amount of ice cream scooped by individuals should be independent.

**Randomization condition:** Subjects were randomly assigned to groups.

**Nearly Normal condition:** Assume that this condition is met.

Since the conditions are satisfied, it is appropriate to model the sampling distribution of the difference in means with a Student's $t$-model, with 34 degrees of freedom (from the approximation formula). Perform a two-sample $t$-test.

The sampling distribution model has mean 0, with standard error:

$$SE(\bar{y}_{big} - \bar{y}_{small}) = \sqrt{\frac{2.91^2}{22} + \frac{1.84^2}{26}} \approx 0.7177 \text{ oz.}$$

The observed difference between the mean amounts is 6.58 – 5.07 = 1.51 oz.

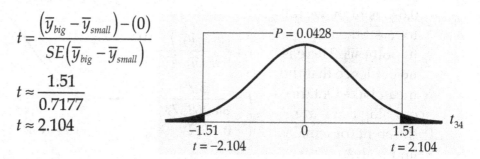

$$t = \frac{\left(\bar{y}_{big} - \bar{y}_{small}\right) - (0)}{SE\left(\bar{y}_{big} - \bar{y}_{small}\right)}$$

$$t \approx \frac{1.51}{0.7177}$$

$$t \approx 2.104$$

Since the $P$-value of 0.0428 is low, we reject the null hypothesis. There is strong evidence that the that the mean amount of ice cream people put into a bowl is related to the size of the bowl. People tend to put more ice cream into the large bowl, on average, than the small bowl.

## 79. Swim the Lake 2016 revisited.

a)

$$\left(\bar{y}_F - \bar{y}_M\right) \pm t^*_{df} \sqrt{\frac{s_F^2}{n_F} + \frac{s_M^2}{n_M}}$$

$$= (1262.08 - 1226.29) \pm t^*_{39.6} \sqrt{\frac{254.701^2}{36} + \frac{368.022^2}{25}} \approx \left(-135.99, 207.57\right)$$

(This interval was constructed using the summary statistics presented. If you used the full data set, the interval will vary slightly due to rounding.)

If the assumptions and conditions are met, we can be 95% confident that the interval –135.99 to 207.57 minutes contains the true difference in mean crossing times between females and males.  Because the interval includes zero, we cannot be confident that there is any difference at all.

**b)** H$_0$: The mean crossing time is the same for females and males.

$$\left(\mu_F = \mu_M \ \text{ or } \ \mu_F - \mu_M = 0\right)$$

H$_A$: The mean crossing time is different for females and males.

$$\left(\mu_F \neq \mu_M \ \text{ or } \ \mu_F - \mu_M \neq 0\right)$$

We will check the conditions in part c, but for now we will model the sampling distribution of the difference in means with a Student's $t$-model, with 39.6 degrees of freedom (from the approximation formula). We will perform a two-sample $t$-test.

The sampling distribution model has mean 0, with standard error:

$$\sqrt{\frac{254.701^2}{36} + \frac{368.022^2}{25}} \approx 84.968373$$

The observed difference between the mean times is 1262.08 – 1226.29 = 35.79 min.

Since the P-value = 0.676 is high, we fail to reject the null hypothesis. There is no evidence that the mean Lake Ontario crossing times are different for females and males.

$$t = \frac{\left(\bar{y}_F - \bar{y}_M\right) - \left(0\right)}{SE\left(\bar{y}_F - \bar{y}_M\right)}$$

$$t \approx \frac{35.79}{84.968373}$$

$$t \approx 0.4212$$

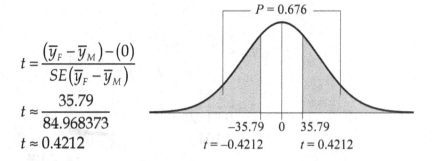

**c)** **Independent groups assumption:** The times from the two groups are likely to be independent of one another, provided that these were all individual swims.
**Randomization condition:** The times are not a random sample from any identifiable population, but it is likely that the times are representative of times from swimmers who might attempt a challenge such as this.  Hopefully, these times were recorded from different swimmers.
**Nearly Normal condition:** The boxplots show two high outliers for the men and some skewness for both. Removing the outliers may make the difference in times larger, but there is no justification for doing so. The histograms are unimodal; but somewhat skewed to the right.

We are reluctant to draw any conclusions about the difference in mean times it takes females and males to swim the lake. The sample is not random, we have no way of knowing if it is representative, and the data are skewed with outliers.

**81. Running heats London.**

$H_0$: The mean time to finish is the same for heats 2 and 5. $\left(\mu_2 = \mu_5 \text{ or } \mu_2 - \mu_5 = 0\right)$

$H_A$: The mean time is not the same for heats 2 and 5. $\left(\mu_2 \neq \mu_5 \text{ or } \mu_2 - \mu_5 \neq 0\right)$

**Independent groups assumption:** The two heats were independent.
**Randomization condition:** Runners were randomly assigned.
**Nearly Normal condition:** The boxplots show an outlier in the distribution of times in heat 2. We will perform the test twice, with and without the outlier.

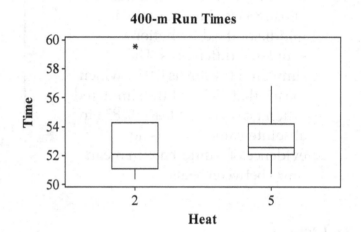

Since the conditions are satisfied, it is appropriate to model the sampling distribution of the difference in means with a Student's *t*-model, with 8.39 degrees of freedom (from the approximation formula). We will perform a two-sample *t*-test.

The sampling distribution model has mean 0, with standard error:

$$SE(\bar{y}_2 - \bar{y}_5) = \sqrt{\frac{3.3618249^2}{6} + \frac{2.0996325^2}{6}} \approx 1.6181.$$

The observed difference between mean times is 53.2267 – 53.1483 = 0.0784.

$$t = \frac{(\bar{y}_2 - \bar{y}_5) - (0)}{SE(\bar{y}_2 - \bar{y}_5)}$$

Since the *P*-value = 0.9625 is high, we fail to reject the null hypothesis. There is no evidence that the mean time to finish differs between the two heats.

$$t \approx \frac{0.0784}{1.6181}$$

$$t \approx 0.048$$

Without the outlier, it is appropriate to model the sampling distribution of the difference in means with a Student's *t*-model, with 8.79 degrees of freedom (from the approximation formula). We will perform a two-sample *t*-test.

The sampling distribution model has mean 0, with standard error:

$$SE(\bar{y}_2 - \bar{y}_5) = \sqrt{\frac{1.4601267^2}{5} + \frac{2.0996325^2}{6}} \approx 1.0776.$$

The observed difference between mean times is 51.962 – 53.1483 = –1.1863.

$$t = \frac{(\bar{y}_2 - \bar{y}_5) - (0)}{SE(\bar{y}_2 - \bar{y}_5)}$$

$$t \approx \frac{-1.1863}{1.0776}$$

$$t \approx -1.10$$

Since the P-value = 0.3001 is high, we fail to reject the null hypothesis. There is no evidence that the mean time to finish differs between the two heats.

Using the randomization test, 500 simulated differences are plotted to the right. The actual difference of 0.07833 is almost exactly in the middle of the distribution of simulated differences. The simulated P-value is 0.952, which means that 95.2% of the simulated differences were at least 0.7833 in absolute value; there is no evidence of a difference in mean times between heats.

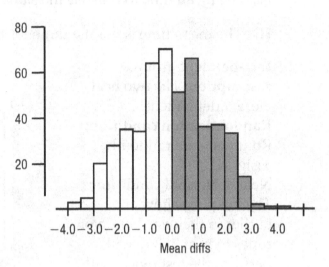

### 83. Tees.

$H_0$: The mean ball velocity is the same for regular and Stinger tees.
$$\left(\mu_S = \mu_R \ \text{ or } \ \mu_S - \mu_R = 0\right)$$

$H_A$: The mean ball velocity is higher for the Stinger tees. $\left(\mu_S > \mu_R \ \text{ or } \ \mu_S - \mu_R > 0\right)$

Assuming the conditions are satisfied, it is appropriate to model the sampling distribution of the difference in means with a Student's *t*-model, with 7.03 degrees of freedom (from the approximation formula). We will perform a two-sample *t*-test.

The sampling distribution model has mean 0, with standard error:

$$SE(\bar{y}_S - \bar{y}_R) = \sqrt{\frac{0.41^2}{6} + \frac{0.89^2}{6}} \approx 0.4000.$$

The observed difference between the mean velocities is 128.83 – 127 = 1.83.

Since the P-value = 0.0013, we reject the null hypothesis. There is strong evidence that the mean ball velocity for stinger tees is higher than the mean velocity for regular tees.

$$t = \frac{(\bar{y}_S - \bar{y}_R) - (0)}{SE(\bar{y}_S - \bar{y}_R)}$$

$$t \approx \frac{1.83}{0.4000}$$

$$t \approx 4.57$$

## 85. Music and memory.

a) $H_0$: The mean memory test score is the same for those who listen to Mozart as it is for those who listen to rap music. $\left(\mu_M = \mu_R \text{ or } \mu_M - \mu_R = 0\right)$

$H_A$: The mean memory test score is greater for those who listen to Mozart than it is for those who listen to rap music. $\left(\mu_M > \mu_R \text{ or } \mu_M - \mu_R > 0\right)$

**Independent groups assumption:** The groups are not related with regards to memory score.
**Randomization condition:** Subjects were randomly assigned to groups.
**Nearly Normal condition:** We don't have the actual data. We will assume that the distributions of the populations of memory test scores are Normal.

Since the conditions are satisfied, it is appropriate to model the sampling distribution of the difference in means with a Student's $t$-model, with 45.88 degrees of freedom (from the approximation formula). We will perform a two-sample $t$-test.

The sampling distribution model has mean 0, with standard error:

$$SE(\overline{y}_M - \overline{y}_R) = \sqrt{\frac{3.19^2}{20} + \frac{3.99^2}{29}} \approx 1.0285 \,.$$

The observed difference between the mean number of objects remembered is

$10.0 - 10.72 = -0.72.$

$$t = \frac{(\overline{y}_M - \overline{y}_R) - (0)}{SE(\overline{y}_M - \overline{y}_R)}$$

$$t \approx \frac{-0.72}{1.0285}$$

$$t \approx -0.70$$

$P = 0.7563$

$-0.72 \quad 0$

$t = -0.70$

$t_{45.88}$

Since the $P$-value = 0.7563 is high, we fail to reject the null hypothesis. There is no evidence that the mean number of objects remembered by those who listen to Mozart is higher than the mean number of objects remembered by those who listen to rap music.

b) $\left(\overline{y}_M - \overline{y}_N\right) \pm t^*_{df} \sqrt{\dfrac{s_M^2}{n_M} + \dfrac{s_N^2}{n_N}} = (10.0 - 12.77) \pm t^*_{19.09} \sqrt{\dfrac{3.19^2}{20} + \dfrac{4.73^2}{13}} \approx \left(-5.351, \, -0.189\right)$

We are 90% confident that the mean number of objects remembered by those who listen to Mozart is between 0.189 and 5.352 objects lower than the mean of those who listened to no music.

## 87. Attendance 2016 revisited.

a) Boxplots of Home Attendance by League are at the right.

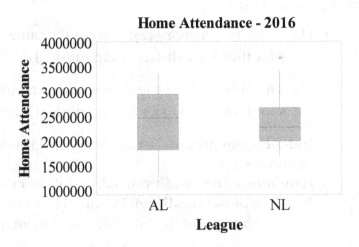

b) A difference of 60,000 is very near the center of the distribution of simulated differences. The vast majority of the simulated differences are more than 60,000 in absolute value, so the difference is not statistically significant.

## 89. Cholesterol and gender II.

Using a randomization test with 500 simulated differences, the actual difference of -16.8 is far from the simulated values. The P-value is 0. In fact, none of the simulated differences are even close to an absolute value of 16.8. This means that there is strong evidence of a difference in mean cholesterol levels between men and women.

# Chapter 18 – Paired Samples and Blocks

## Section 18.1

**1. Which method?**

a) Paired. Each individual has two scores, so they are certainly associated.

b) Not paired. The scores of males and females are independent.

c) Paired. Each student was surveyed twice, so there responses are associated.

## Section 18.2

**3. Cars and trucks.**

a) No. The vehicles have no natural pairing.

b) Possibly. The data are quantitative and paired by vehicle.

c) The sample size is large, but there is at least one extreme outlier that should be investigated before applying these methods.

## Section 18.3

**5. Cars and trucks again.**

$$\bar{d} \pm t_{n-1}^* \left( \frac{s_d}{\sqrt{n}} \right) = 7.37 \pm t_{631}^* \left( \frac{2.52}{\sqrt{632}} \right) \approx (7.17, 7.57)$$

We are 95% confident that the interval 7.17 mpg to 7.57 mpg captures the true improvements in highway gas mileage compared to city gas mileage.

## Section 18.4

**7. Blocking cars and trucks.**

The difference between fuel efficiency of cars and that of trucks can be large, but isn't relevant to the question asked about highway vs. city driving. Pairing places each vehicle in its own block to remove that variation from consideration.

## Chapter Exercises

**9. More eggs?**

a) Randomly assign 50 hens to each of the two kinds of feed. Compare the mean egg production of the two groups at the end of one month.

**b)** Randomly divide the 100 hens into two groups of 50 hens each. Feed the hens in the first group the regular feed for two weeks, then switch to the additive for 2 weeks. Feed the hens in the second group the additive for two weeks, and then switch to the regular feed for two weeks. Subtract each hen's "regular" egg production from her "additive" egg production, and analyze the mean difference in egg production.

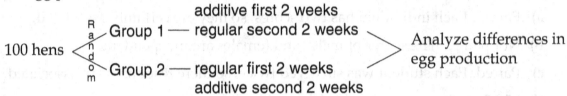

**c)** The matched pairs design in part b is the stronger design. Hens vary in their egg production regardless of feed. This design controls for that variability by matching the hens with themselves.

## 11. Sex sells?

**a)** Randomly assign half of the volunteers to watch ads with sexual images, and assign the other half to watch ads without the sexual images. Record the number of items remembered. Then have each group watch the other type of ad. Record the number of items recalled. Examine the difference in the number of items remembered for each person.

Volunteers — Randomdom — Group 1 — Ads with sexual images first — Analyze mean difference in number of items remembered
Group 2 — Ads with no sexual images first

**b)** Randomly assign half of the volunteers to watch ads with sexual images, and assign the other half to watch ads without the sexual images. Record the number of items remembered. Compare the mean number of products remembered by each group.

Volunteers — Randomdom — Group 1 — Ads with sexual images — Compare mean number of items remembered
Group 2 — Ads with no sexual images

## 13. Women.

**a)** The paired *t*-test is appropriate. The labor force participation rate for two different years was paired by city.

**b)** Since the *P*-value = 0.0244, there is evidence of a difference in the average labor force participation rate for women between 1968 and 1972. The evidence suggests an increase in the participation rate for women.

**15. Friday the 13th, traffic.**

a) The paired *t*-test is appropriate, since we have 10 pairs of Fridays in 5 different months. Data from adjacent Fridays within a month may be more similar than randomly chosen Fridays.

b) Since the *P*-value = 0.0004, there is evidence that the mean number of cars on the M25 motorway on Friday the 13th is less than the mean number of cars on the previous Friday.

c) We don't know if these Friday pairs were selected at random. The Nearly Normal condition appears to be met by the differences, but the sample size of ten pairs is small. Additionally, mixing data collected at both Junctions 9 and 10 together in one test confounds the test. We don't know if we are seeing a Friday the 13th effect or a Junction effect.

**17. Online insurance I.**

Adding variances requires that the variables be independent. These price quotes are for the same cars, so they are paired. Drivers quoted high insurance premiums by the local company will be likely to get a high rate from the online company, too.

**19. Online insurance II.**

a) The histogram would help you decide whether the online company offers cheaper insurance. We are concerned with the difference in price, not the distribution of each set of prices.

b) Insurance cost is based on risk, so drivers are likely to see similar quotes from each company, making the differences relatively smaller.

c) The price quotes are paired. They were for a random sample of the agent's customers and the histogram of differences looks approximately Normal.

**21. Online insurance III.**

$H_0$: The mean difference between online and local insurance rates is zero. $\left(\mu_{Local-Online} = 0\right)$

$H_A$: The mean difference is greater than zero. $\left(\mu_{Local-Online} > 0\right)$

Since the conditions are satisfied (in a previous exercise), the sampling distribution of the difference can be modeled with a Student's *t*-model with 10 – 1 = 9 degrees of freedom, $t_9\left(0, \dfrac{175.663}{\sqrt{10}}\right)$.

We will use a paired *t*-test, with $\bar{d} = 45.9$.

Since the *P*-value = 0.215 is high, we fail to reject the null hypothesis. There is no evidence that online insurance premiums are lower on average.

$$t = \frac{\bar{d} - 0}{\frac{s}{\sqrt{n}}}$$

$$t = \frac{45.9 - 0}{\frac{175.663}{\sqrt{10}}}$$

$$t \approx 0.83$$

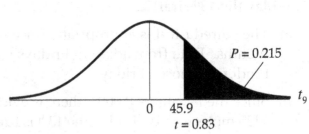

$P = 0.215$

$t_9$

0   45.9

$t = 0.83$

### 23. City Temperatures.

a) **Paired data assumption:** The data are paired by city.
**Randomization condition:** These cities might not be representative of all cities, so be cautious in generalizing the results.
**Normal population assumption:** The histogram of differences between January and July mean temperature is roughly unimodal and symmetric, but

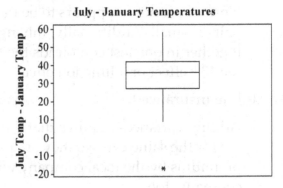

shows a low outlier. This is Auckland, New Zealand, a city in the southern hemisphere. Seasons here would be the opposite of the rest of the cities, which are in the northern hemisphere. It should be set aside.

b) With Auckland set aside, the conditions are satisfied. The sampling distribution of the difference can be modeled with a Student's *t*-model with 11 − 1 = 10 degrees of freedom. We will find a paired *t*-interval, with 95% confidence.

$$\bar{d} \pm t_{n-1}^* \left( \frac{s_d}{\sqrt{n}} \right) = 33.7429 \pm t_{34}^* \left( \frac{14.9968}{\sqrt{34}} \right) \approx (30.94, 39.53)$$

We are 95% confident that the average high temperature in northern hemisphere cities in July is an average of between 30.94° to 39.53° higher than in January.

c) Answers will vary. Based on our simulation, we are 95% confident that the average high temperature in northern hemisphere cities in July is an average of between 28.9° to 38.1° higher than in January.

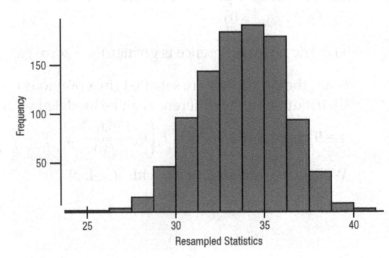

## 25. Push-ups.

**Independent groups assumption:** The group of boys is independent of the group of girls.

**Randomization condition:** Assume that students are assigned to gym classes at random.

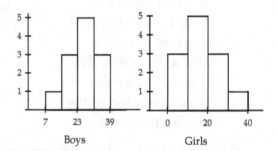

Boys                    Girls

**Nearly Normal condition:** The histograms of the number of push-ups from each group are roughly unimodal and symmetric.

Since the conditions are satisfied, it is appropriate to model the sampling distribution of the difference in means with a Student's *t*-model, with 21 degrees of freedom (from the approximation formula). We will construct a two-sample *t*-interval, with 90% confidence.

$$(\bar{y}_B - \bar{y}_G) \pm t^*_{df} \sqrt{\frac{s_B^2}{n_B} + \frac{s_G^2}{n_G}} = (23.8333 - 16.5000) \pm t^*_{21} \sqrt{\frac{7.20900^2}{12} + \frac{8.93919^2}{12}} \approx (1.6, 13.0)$$

We are 90% confident that, at Gossett High, the mean number of push-ups that boys can do is between 1.6 and 13.0 more than the mean for the girls.

## 27. Job satisfaction.

**a)** Use a paired *t*-test.

**Paired data assumption:** The data are before and after job satisfaction rating for the same workers.

**Randomization condition:** The workers were randomly selected to participate.

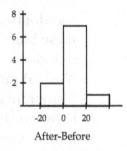

After-Before

**Nearly Normal conditon:** The histogram of differences between before and after job satisfaction ratings is roughly unimodal and symmetric.

**b)** H₀: The mean difference in before and after job satisfaction scores is zero, and the exercise program is not effective at improving job satisfaction. $(\mu_d = 0)$

Hₐ: The mean difference is greater than zero, and the exercise program is effective at improving job satisfaction. $(\mu_d > 0)$

Since the conditions are satisfied, the sampling distribution of the difference can be modeled with a Student's *t*-model with $10 - 1 = 9$ degrees of freedom,

$t_9 \left( 0, \dfrac{7.47217}{\sqrt{10}} \right)$. We will use a paired *t*-test, with $\bar{d} = 8.5$.

Since the *P*-value = 0.0029 is low, we reject the null hypothesis. There is evidence that the mean job satisfaction rating has increased since the implementation of the exercise program.

$$t = \frac{\bar{d} - 0}{\frac{s_d}{\sqrt{n}}}$$

$$t = \frac{8.5 - 0}{\frac{7.47217}{\sqrt{10}}}$$

$$t \approx 3.60$$

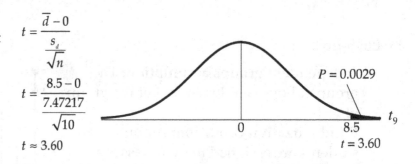

$P = 0.0029$

$t_9$

0

8.5

$t = 3.60$

c) We concluded that there was an increase job satisfaction rating. If we are wrong, and there actually was no increase, we have committed a Type I error.

d) The *P*-value of the sign test is 0.1094, which would lead us to fail to reject the null hypothesis that the median difference was 0. This is a different conclusion than the paired *t*-test. However, since the conditions for the paired *t*-test are met, we should use those results. The *t*-test has more power.

**29. Yogurt.**

H₀: The mean difference in calories between servings of strawberry and vanilla yogurt is zero. $(\mu_d = 0)$

Hₐ: The mean difference in calories between servings of strawberry and vanilla yogurt is different from zero. $(\mu_d \neq 0)$

**Paired data assumption:** The yogurt is paired by brand.
**Randomization condition:** Assume that these brands are representative of all brands.
**Normal population assumption:** The histogram of differences in calorie content between strawberry and vanilla shows an outlier, Great Value. When the outlier is eliminated, the histogram of differences is roughly unimodal and symmetric.

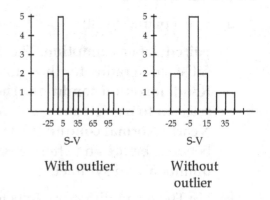

S-V

With outlier

S-V

Without outlier

When Great Value yogurt is removed, the conditions are satisfied. The sampling distribution of the difference can be modeled with a Student's *t*-model with

$11 - 1 = 10$ degrees of freedom, $t_{10}\left(0, \dfrac{18.0907}{\sqrt{11}}\right)$.

We will use a paired *t*-test, with $\bar{d} \approx 4.54545$.

Since the *P*-value = 0.4241 is high, we fail to reject the null hypothesis. There is no evidence of a mean difference in calorie content between strawberry yogurt and vanilla yogurt.

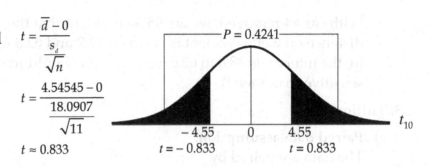

$$t = \frac{\bar{d} - 0}{\frac{s_d}{\sqrt{n}}}$$

$$t = \frac{4.54545 - 0}{\frac{18.0907}{\sqrt{11}}}$$

$$t \approx 0.833$$

**31. Stopping distance.**

a) **Randomization Condition:** These cars are not a random sample, but are probably representative of all cars in terms of stopping distance.
**Nearly Normal Condition:** A histogram of the stopping distances is skewed to the right, but this may just be sampling variation from a Normal population. The "skew" is only a couple of stopping distances. We will proceed cautiously.

The cars in the sample had a mean stopping distance of 138.7 feet and a standard deviation of 9.66149 feet. Since the conditions have been satisfied, construct a one-sample *t*-interval, with 10 – 1 = 9 degrees of freedom, at 95% confidence.

$$\bar{y} \pm t^*_{n-1} \left( \frac{s}{\sqrt{n}} \right) = 138.7 \pm t^*_9 \left( \frac{9.66149}{\sqrt{10}} \right) \approx (131.8,\ 145.6)$$

We are 95% confident that the mean dry pavement stopping distance for cars with this type of tires is between 131.8 and 145.6 feet. This estimate is based on an assumption that these cars are representative of all cars and that the population of stopping distances is Normal.

b) **Paired data assumption:** The data are paired by car.
**Randomization condition:** Assume that the cars are representative of all cars.
**Normal population assumption:** The difference in stopping distance for car #4 is an outlier, at only 12 feet. After excluding this difference, the histogram of differences is unimodal and symmetric.

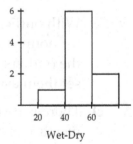

Wet-Dry

Since the conditions are satisfied, the sampling distribution of the difference can be modeled with a Student's *t*-model with 9 – 1 = 8 degrees of freedom. We will find a paired *t*-interval, with 95% confidence.

$$\bar{d} \pm t^*_{n-1} \left( \frac{s_d}{\sqrt{n}} \right) = 55 \pm t^*_8 \left( \frac{10.2103}{\sqrt{9}} \right) \approx (47.2,\ 62.8)$$

With car #4 removed, we are 95% confident that the mean increase in stopping distance on wet pavement is between 47.2 and 62.8 feet. (If you leave the outlier in, the interval is 38.8 to 62.6 feet, but you should remove it! This procedure is sensitive to outliers!)

**33. Tuition 2016.**

a) **Paired data assumption:** The data are paired by college.
**Randomization condition:** The colleges were selected randomly.
**Normal population assumption:** The tuition difference for UC Irvine, at $26,682, is an outlier, as is the tuition difference for New College of Florida at $23,078. Once these have

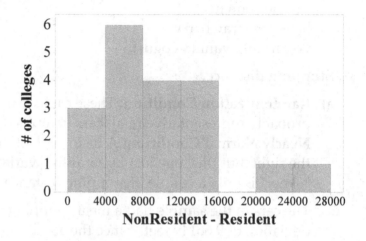

been set aside, the histogram of the differences is roughly unimodal and slightly skewed to the right. This should be fine for inference in a sample of 17 colleges.

b) Since the conditions are satisfied, the sampling distribution of the difference can be modeled with a Student's *t*-model with 17 – 1 = 16 degrees of freedom. We will find a paired *t*-interval, with 90% confidence.

$$\bar{d} \pm t^*_{n-1}\left(\frac{s_d}{\sqrt{n}}\right) = 8158.5882 \pm t^*_{16}\left(\frac{3968.3927}{\sqrt{17}}\right) \approx (6478.216, \ 9838.960)$$

With outliers removed, we are 90% confident that the mean increase in tuition for nonresidents versus residents is between about $6478 and $9839. (If you left the outliers in your data, the interval is about $7335 to $12,501, but you should set them aside! This procedure is sensitive to the presence of outliers!)

c) There is no evidence to suggest that the magazine made a false claim. An increase of $7000 for nonresidents is contained within our 90% confidence interval. (If the outliers were left in, then $7000 is actually lower than the interval of plausible values for the mean tuition difference.)

**35. Strikes.**

a) Since 60% of 50 pitches is 30 pitches, the Little Leaguers would have to throw an average of more than 30 strikes in order to give support to the claim made by the advertisements.

H₀: The mean number of strikes thrown by Little Leaguers who have completed the training is 30. $(\mu_A = 30)$

Hₐ: The mean number of strikes thrown by Little Leaguers who have completed the training is greater than 30. $(\mu_A > 30)$

**Randomization Condition:** Assume that these players are representative of all Little League pitchers.
**Nearly Normal Condition:** The histogram of the number of strikes thrown after the training is roughly unimodal and symmetric.

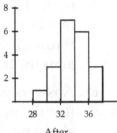

After

The pitchers in the sample threw a mean of 33.15 strikes, with a standard deviation of 2.32322 strikes. Since the conditions for inference are satisfied, we can model the sampling distribution of the mean number of strikes thrown with a Student's $t$ model, with 20 – 1 = 19 degrees of freedom, $t_{19}\left(30, \dfrac{2.32322}{\sqrt{20}}\right)$.

We will perform a one-sample $t$-test.

Since the $P$-value = $3.92 \times 10^{-6}$ is very low, we reject the null hypothesis. There is strong evidence that the mean number of strikes that Little Leaguers can throw after the training is more than 30. (This test says nothing about the effectiveness of the training; just that Little Leaguers can throw more than 60% strikes on average after completing the training. This might not be an improvement.)

$$t = \frac{\bar{y}_A - \mu_0}{\dfrac{s_A}{\sqrt{n_A}}}$$

$$t = \frac{33.15 - 30}{\dfrac{2.32322}{\sqrt{20}}}$$

$$t = 6.06$$

**b)** H₀: The mean difference in number of strikes thrown before and after the training is zero. $(\mu_d = 0)$

Hₐ: The mean difference in number of strikes thrown before and after the training is greater than zero. $(\mu_d > 0)$

**Paired data assumption:** The data are paired by pitcher.
**Randomization condition:** Assume that these players are representative of all Little League pitchers.
**Normal population assumption:** The histogram of differences is roughly unimodal and symmetric.

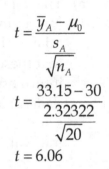

After-Before

Since the conditions are satisfied, the sampling distribution of the difference can be modeled with a Student's $t$-model with 20 – 1 = 19 degrees of freedom, $t_{19}\left(0, \dfrac{3.32297}{\sqrt{19}}\right)$.

We will use a paired $t$-test, with $\bar{d} = 0.1$.

Since the $P$-value = 0.4472 is high, we fail to reject the null hypothesis. There is no evidence of a mean difference in number of strikes thrown before and

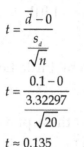

$$t = \frac{\bar{d} - 0}{\frac{s_d}{\sqrt{n}}}$$

$$t = \frac{0.1 - 0}{\frac{3.32297}{\sqrt{20}}}$$

$$t \approx 0.135$$

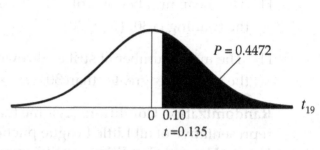

after the training. The training does not appear to be effective.

c) Answers will vary. According to our bootstrap interval, we are 95% confident that the mean number of strikes thrown is between 32.2 and 34.5. This is well above the 30 strikes out of 50 pitches (60%) claimed by the advertisements. There is evidence that Little Leaguers can throw at least 60% strikes after the training program.

**37. Wheelchair marathon 2016.**

a) The data are certainly paired. Even if the individual times show a trend of improving speed over time, the differences may well be independent of each other. They are subject to random year-to-year fluctuations, and we may believe that these data are representative of similar races. We don't have any information with which to check the Nearly Normal condition, unless we use the data set instead of the summary statistics provided. At any rate, with a sample of size 40, the Central Limit Theorem will allow us to construct the confidence interval.

b) $\bar{d} \pm t^*_{n-1} \left( \frac{s_d}{\sqrt{n}} \right) = -7.27 \pm t^*_{39} \left( \frac{33.568}{40} \right) \approx (-18.01, 3.47)$

We are 95% confident that the interval –18.01 to 3.47 minutes contains the true mean time difference for women's wheelchair times and men's running times.

c) The interval contains zero, so we would not reject a null hypothesis of no mean difference at a significance level of 0.05. We are unable to discern a difference between the female wheelchair times and the male running times.

**39. BST.**

a) **Paired data assumption:** We are testing the same cows, before and after injections of BST.
**Randomization condition:** These cows are likely to be representative of all Ayrshires.
**Normal population assumption:** We don't have the list of individual differences, so we can't look at a histogram. The sample is large, so we may proceed.

Since the conditions are satisfied, the sampling distribution of the difference can be modeled with a Student's *t*-model with 60 – 1 = 59 degrees of freedom. We will find a paired *t*-interval, with 95% confidence.

b) $\bar{d} \pm t^*_{n-1}\left(\dfrac{s_d}{\sqrt{n}}\right) = 14 \pm t^*_{59}\left(\dfrac{5.2}{\sqrt{60}}\right) \approx (12.66, 15.34)$

c) We are 95% confident that the mean increase in daily milk production for Ayshire cows after BST injection is between 12.66 and 15.34 pounds.

d) 25% of 47 pounds is 11.75 pounds. According to the interval generated in part b, the average increase in milk production is more than this, so the farmer can justify the extra expense for BST.

# Chapter 19 – Comparing Counts

## Section 19.1

### 1. Human births.

a) If there were no "seasonal effect" we would expect 25% of births to occur in each season. The expected number of births is 0.25(120) = 30 births per season.

b) $\chi^2 = \dfrac{(25-30)^2}{30} + \dfrac{(35-30)^2}{30} + \dfrac{(32-30)^2}{30} + \dfrac{(28-30)^2}{30} \approx 1.933$

c) There are 4 seasons, so there are 4 – 1 = 3 degrees of freedom.

### 3. Human births, again.

a) The mean of the $\chi^2$ distribution is the number of degrees of freedom, so we would expect the $\chi^2$ statistic to be 3 if there were no seasonal effect.

b) Since 1.933 is less than the mean of 3, it does not seem large in comparison.

c) We should fail to reject the null hypothesis. These data do not provide evidence of a seasonal effect on human births.

d) The critical value of $\chi^2$ with 3 degrees of freedom and $\alpha = 0.05$ is 7.815.

e) Since $\chi^2 = 1.933$ is less than the critical value, 7.815, we fail to reject the null hypothesis. There is no evidence of a seasonal effect on human births.

## Section 19.2

### 5. Customer ages.

a) The null hypothesis is that the age distributions of the customers are the same at the two branches.

b) There are three age groups and one variable, location of the branch. This is a $\chi^2$ test of homogeneity.

c) Expected Counts

| | \multicolumn{3}{c}{Age} | |
|---|---|---|---|---|
| | Less than 30 | 30-55 | 56 or older | Total |
| In-Town Branch | 25 | 45 | 30 | 100 |
| Mall Branch | 25 | 45 | 30 | 100 |
| Total | 50 | 90 | 60 | 200 |

d) $\chi^2 = \dfrac{(20-25)^2}{25} + \dfrac{(40-45)^2}{45} + \dfrac{(40-30)^2}{30} + \dfrac{(30-25)^2}{25} + \dfrac{(50-45)^2}{45} + \dfrac{(20-30)^2}{30} \approx 9.778$

e) There are 2 rows and 3 columns, so there are (2 – 1)(3 – 1) = 2 degrees of freedom.

**f)** The probability of having a $\chi^2$ value over 9.778 with df = 2 is 0.0075.

**g)** Since the P-value is so low, reject the null hypothesis. There is strong evidence that the distribution of ages is not the same at the two branches. The mall branch had more customers under 30, and fewer customers 56 or older, than expected. The in-town branch had more customers 56 and older, and fewer customers under 30, than expected.

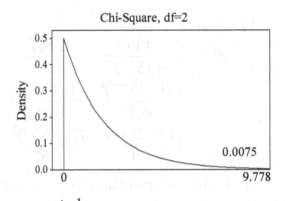

## Section 19.3

**7. Human births, last time.**

**a)** The standardized residuals are:

$$c_1 = \frac{(25-30)}{\sqrt{30}} \approx -0.913 \qquad c_2 = \frac{(35-30)}{\sqrt{30}} \approx 0.913$$

$$c_3 = \frac{(32-30)}{\sqrt{30}} \approx 0.365 \qquad c_4 = \frac{(28-30)}{\sqrt{30}} \approx -0.365$$

**b)** None of the standardized residuals are large. Since they are z-scores, they are actually quite small.

**c)** We did not reject the null hypothesis, so we should expect the standardized residuals to be relatively small.

## Section 19.4

**9. *Iliad* injuries 800 BCE**

**a)** The null hypothesis is that the lethality of an injury is independent of the injury site.

| Expected Counts | | Lethal? | | |
|---|---|---|---|---|
| | | Fatal | Not fatal | Total |
| Injury Site | body | 54.15 | 12.85 | 67 |
| | head/neck | 36.37 | 8.63 | 45 |
| | limb | 27.48 | 6.52 | 34 |
| | Total | 118 | 28 | 1467 |

b)

$$\chi^2 = \frac{(61-54.15)^2}{54.15} + \frac{(6-12.85)^2}{12.85}$$
$$+ \frac{(44-36.37)^2}{36.37} + \frac{(1-8.63)^2}{8.63}$$
$$+ \frac{(13-27.48)^2}{27.48} + \frac{(21-6.52)^2}{6.52} \approx 52.65$$

c) There are 3 rows and 2 columns, so there are $(3-1)(2-1) = 2$ degrees of freedom.

d) The probability of having a $\chi^2$ value over 52.65 with df = 2 is less than 0.0001.

e) Since the P-value is low, reject the null hypothesis. There is evidence of an association between the site of the injury and whether or not the injury was lethal. Injuries to the head/neck and body are more likely to be fatal, while injuries to the limbs are less likely to be fatal.

**Chapter Exercises.**

**11. Which test?**

a) Chi-square test of Independence. We have one sample and two variables. We want to see if the variable *account type* is independent of the variable *trade type*.

b) Some other statistics test. The variable *account size* is quantitative, not counts.

c) Chi-square test of Homogeneity. We have two samples (residential and non-residential students), and one variable, *courses*. We want to see if the distribution of *courses* is the same for the two groups.

**13. Dice.**

a) If the die were fair, you'd expect each face to show 10 times.

b) Use a chi-square test for goodness-of-fit. We are comparing the distribution of a single variable (outcome of a die roll) to an expected distribution.

c) $H_0$: The die is fair. (All faces have the same probability of coming up.)

$H_A$: The die is not fair. (Some faces are more or less likely to come up than others.)

d) **Counted data condition:** We are counting the number of times each face comes up.
**Randomization condition:** Die rolls are random and independent of each other.
**Expected cell frequency condition:** We expect each face to come up 10 times, and 10 is greater than 5.

e) Under these conditions, the sampling distribution of the test statistic is $\chi^2$ on $6 - 1 = 5$ degrees of freedom. We will use a chi-square goodness-of-fit test.

**f)**

| Face | Observed | Expected | Residual = $(Obs - Exp)$ | $(Obs - Exp)^2$ | Component = $\dfrac{(Obs - Exp)^2}{Exp}$ |
|------|----------|----------|--------------------------|-----------------|------------------------------------------|
| 1 | 11 | 10 | 1 | 1 | 0.1 |
| 2 | 7 | 10 | – 3 | 9 | 0.9 |
| 3 | 9 | 10 | – 1 | 1 | 0.1 |
| 4 | 15 | 10 | 5 | 25 | 2.5 |
| 5 | 12 | 10 | 2 | 4 | 0.4 |
| 6 | 6 | 10 | – 4 | 16 | 1.6 |

$$\sum = 5.6$$

**g)** Since the *P*-value = 0.3471 is high, we fail to reject the null hypothesis. There is no evidence that the die is unfair.

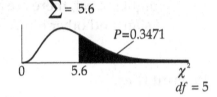

## 15. Nuts.

**a)** The weights of the nuts are quantitative. Chi-square goodness-of-fit requires counts.

**b)** In order to use a chi-square test, you could count the number of each type of nut. However, it's not clear whether the company's claim was a percentage by number or a percentage by weight.

## 17. NYPD and race.

$H_0$: The distribution of ethnicities in the police department represents the distribution of ethnicities of the youth of New York City.

$H_A$: The distribution of ethnicities in the police department does not represent the distribution of ethnicities of the youth of New York City.

**Counted data condition:** The percentages reported must be converted to counts.
**Randomization condition:** Assume that the current NYPD is representative of recent departments with respect to ethnicity.
**Expected cell frequency condition:** The expected counts are all much greater than 5.

(Note: The observed counts should be whole numbers. They are actual policemen. The expected counts may be decimals, since they behave like averages.)

| Ethnicity | Observed | Expected |
|-----------|----------|----------|
| White | 16965 | 7644.852 |
| Black | 3796 | 7383.042 |
| Latino | 5001 | 8247.015 |
| Asian | 367 | 2382.471 |
| Other | 52 | 523.620 |

Under these conditions, the sampling distribution of the test statistic is $\chi^2$ on 5 – 1 = 4 degrees of freedom. We will use a chi-square goodness-of-fit test.

$$\chi^2 = \sum_{all\,cells} \frac{(Obs - Exp)^2}{Exp} = \frac{(16965 - 7644.852)^2}{7644.852} + \frac{(3796 - 7383.042)^2}{7383.042} + \frac{(5001 - 8247.015)^2}{8247.015}$$

$$+ \frac{(367 - 2382.471)^2}{2382.471} + \frac{(52 - 523.620)^2}{523.620} \approx 16,500$$

With $\chi^2$ of over 16,500, on 4 degrees of freedom, the *P*-value is essentially 0, so we reject the null hypothesis. There is strong evidence that the distribution of ethnicities of NYPD officers does not represent the distribution of ethnicities of the youth of New York City. Specifically, the proportion of white officers is much higher than the proportion of white youth in the community. As one might expect, there are also lower proportions of officers who are black, Latino, Asian, and other ethnicities than we see in the youth in the community.

**19. Fruit flies.**

**a)** $H_0$: The ratio of traits in this type of fruit fly is 9:3:3:1, as genetic theory predicts.

$H_A$: The ratio of traits in this type of fruit fly is not 9:3:3:1.

**Counted data condition:** The data are counts.
**Randomization condition:** Assume that these flies are representative of all fruit flies of this type.
**Expected cell frequency condition:** The expected counts are all greater than 5.

| trait | Observed | Expected |
|-------|----------|----------|
| YN | 59 | 56.25 |
| YS | 20 | 18.75 |
| EN | 11 | 18.75 |
| ES | 10 | 6.25 |

Under these conditions, the sampling distribution of the test statistic is $\chi^2$ on $4 - 1 = 3$ degrees of freedom. We will use a chi-square goodness-of-fit test.

$$\chi^2 = \sum_{all\,cells} \frac{(Obs - Exp)^2}{Exp} = \frac{(59 - 56.25)^2}{56.25} + \frac{(20 - 18.75)^2}{18.75} + \frac{(11 - 18.75)^2}{18.75} + \frac{(10 - 6.25)^2}{6.25} \approx 5.671$$

With $\chi^2 \approx 5.671$, on 3 degrees of freedom, the *P*-value = 0.1288 is high, so we fail to reject the null hypothesis. There is no evidence that the ratio of traits is different than the theoretical ratio predicted by the genetic model. The observed results are consistent with the genetic model.

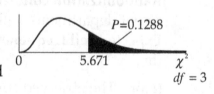

**b)** With $\chi^2 \approx 11.342$, on 3 degrees of freedom, the P-value = 0.0100 is low, so we reject the null hypothesis. There is strong evidence that the ratio of traits is different than the theoretical ratio predicted by the genetic model. Specifically, there is evidence that the normal wing length occurs less frequently than expected and the short wing length occurs more frequently than expected.

| trait | Observed | Expected |
|-------|----------|----------|
| YN | 118 | 112.5 |
| YS | 40 | 37.5 |
| EN | 22 | 37.5 |
| ES | 20 | 12.5 |

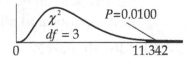

**c)** At first, this seems like a contradiction. We have two samples with exactly the same ratio of traits. The smaller of the two provides no evidence of a difference, yet the larger one provides strong evidence of a difference. This is explained by the sample size. In general, large samples decrease the proportion of variation from the true ratio. Because of the relatively small sample in the first test, we are unwilling to say that there is a difference. There just isn't enough evidence. But the larger sample allows us to be more certain about the difference.

**21. Hurricane frequencies.**

**a)** We would expect 96/16 = 6 hurricanes per time period.

**b)** We are comparing the distribution of the number of hurricanes, a single variable, to a theoretical distribution. A Chi-square test for goodness-of-fit is appropriate.

**c)** $H_0$: The number of large hurricanes remains constant over decades.

$H_A$: The number of large hurricanes has changed.

**d)** There are 16 time periods, so there are 16 – 1 = 15 degrees of freedom.

**e)** $P(\chi^2_{df=15} > 12.67) \approx 0.63$

**f)** The very high P-value means that these data offer no evidence that the number of hurricanes large hurricanes has changed.

**23. Childbirth, part 1.**

**a)** There are two variables, breastfeeding and having an epidural, from a single group of births. We will perform a Chi-square test for Independence.

**b)** $H_0$: Breastfeeding success is independent of having an epidural.

$H_A$: There is an association between breastfeeding success and having an epidural.

**25. Childbirth, part 2.**

**a)** The table has 2 rows and 2 columns, so there are $(2-1)\times(2-1)=1$ degree of freedom.

b) We expect $\dfrac{474}{1178} \approx 40.2\%$ of all babies to not be breastfeeding after 6 months, so we expect that 40.2% of the 396 epidural babies, or 159.34, to not be breastfeeding after 6 months.

c) Breastfeeding behavior should be independent for these babies. They are fewer than 10% of all babies, and we assume they are representative of all babies. We have counts, and all the expected cells are at least 5.

## 27. Childbirth, part 3.

a) $\dfrac{(Obs - Exp)^2}{Exp} = \dfrac{(190 - 159.34)^2}{159.34} = 5.90$   b) $P(\chi^2_{df=1} > 14.87) < 0.005$

c) The $P$-value is very low, so reject the null hypothesis. There's strong evidence of an association between having an epidural and subsequent success in breastfeeding.

## 29. Childbirth, part 4.

a) $c = \dfrac{Obs - Exp}{\sqrt{Exp}} = \dfrac{190 - 159.34}{\sqrt{159.34}} = 2.43$

b) It appears that babies whose mothers had epidurals during childbirth are much more likely to be breastfeeding 6 months later.

## 31. Childbirth, part 5.

These factors would not have been mutually exclusive. There would be yes or no responses for every baby for each.

## 33. Titanic.

a) $P(\text{crew}) = \dfrac{889}{2208} \approx 0.4026$   b) $P(\text{third and alive}) = \dfrac{180}{2208} \approx 0.0815$

c) $P(\text{alive} \mid \text{first}) = \dfrac{P(\text{alive and first})}{P(\text{first})} = \dfrac{201/2208}{324/2208} = \dfrac{201}{324} \approx 0.6204$

d) The overall chance of survival is $\dfrac{712}{2208} \approx 0.3225$, so we would expect about 32.3% of the crew, or about 286.67 members of the crew, to survive.

e) $H_0$: Survival was independent of status on the ship.

   $H_A$: Survival depended on status on the ship.

f) The table has 2 rows and 4 columns, so there are $(2-1) \times (4-1) = 3$ degrees of freedom.

**g)** With $\chi^2 \approx 187.56$, on 3 degrees of freedom, the $P$-value is essentially 0, so we reject the null hypothesis. There is strong evidence of an association between survival and status. First-class passengers were more likely to survive than any other class or crew.

## 35. Titanic again.

First class passengers were most likely to survive, while third class passengers and crew were under-represented among the survivors.

## 37. Cranberry juice.

**a)** This is an experiment. Volunteers were assigned to drink a different beverage.

**b)** We are concerned with the proportion of urinary tract infections among three different groups. We will use a chi-square test for homogeneity.

**c)** $H_0$: The proportion of urinary tract infection is the same for each group.

$H_A$: The proportion of urinary tract infection is different among the groups.

**d)** **Counted data condition:** The data are counts.
**Randomization condition:** Although not specifically stated, we will assume that the women were randomly assigned to treatments.
**Expected cell frequency condition:** The expected counts are all greater than 5.

|  | Cranberry (Obs/Exp) | Lactobacillus (Obs/Exp) | Control (Obs/Exp) |
|---|---|---|---|
| **Infection** | 8 / 15.333 | 20 / 15.333 | 18 / 15.333 |
| **No infection** | 42 / 34.667 | 30 / 34.667 | 32 / 34.667 |

**e)** The table has 2 rows and 3 columns, so there are $(2-1)\times(3-1)=2$ degrees of freedom.

**f)** $\chi^2 = \sum_{all\,cells} \dfrac{(Obs - Exp)^2}{Exp} \approx 7.776$

$P$-value $\approx 0.020$.

**g)** Since the $P$-value is low, we reject the null hypothesis. There is strong evidence of difference in the proportion of urinary tract infections for cranberry juice drinkers, lactobacillus drinkers, and women that drink neither of the two beverages.

**h)** A table of the standardized residuals is below, calculated by using $c = \dfrac{Obs - Exp}{\sqrt{Exp}}$.

|  | Cranberry | Lactobacillus | Control |
|---|---|---|---|
| Infection | –1.87276 | 1.191759 | 0.681005 |
| No infection | 1.245505 | –0.79259 | –0.45291 |

There is evidence that women who drink cranberry juice are less likely to develop urinary tract infections, and women who drank lactobacillus are more likely to develop urinary tract infections.

**39. Montana.**

**a)** We have one group, categorized according to two variables, political party and being male or female, so we will perform a chi-square test for independence.

**b)** $H_0$: Political party is independent of being male or female in Montana.

$H_A$: There is an association between political party and being male or female in Montana.

**c)** **Counted data condition:** The data are counts.
**Randomization condition:** Although not specifically stated, we will assume that the poll was conducted randomly.
**Expected cell frequency condition:** The expected counts are all greater than 5.

|  | Democrat (Obs / Exp) | Republican (Obs / Exp) | Independent (Obs / Exp) |
|---|---|---|---|
| Male | 36 / 43.663 | 45 / 40.545 | 24 / 20.792 |
| Female | 48 / 40.337 | 33 / 37.455 | 16 / 19.208 |

**d)** Under these conditions, the sampling distribution of the test statistic is $\chi^2$ on 2 degrees of freedom. We will use a chi-square test for independence.

$$\chi^2 = \sum_{all\,cells} \frac{(Obs - Exp)^2}{Exp} \approx 4.851$$

The *P*-value ≈ 0.0884

P=0.0884

$\chi^2$
df = 2

0    4.851

**e)** Since the *P*-value ≈ 0.0884 is fairly high, we fail to reject the null hypothesis. There is little evidence of an association between being male or female and political party in Montana.

**41. Montana revisited.**

$H_0$: Political party is independent of region in Montana.

$H_A$: There is an association between political party and region in Montana.

**Counted data condition:** The data are counts.
**Randomization condition:** Although not specifically stated, we will assume that the poll was conducted randomly.

**Expected cell frequency condition:** All expected counts are greater than 5.

|  | Democrat (Obs / Exp) | Republican (Obs / Exp) | Independent (Obs / Exp) |
|---|---|---|---|
| West | 39 / 28.277 | 17 / 26.257 | 12 / 13.465 |
| Northeast | 15 / 23.703 | 30 / 22.01 | 12 / 11.287 |
| Southeast | 30 / 32.02 | 31 / 29.733 | 16 / 15.248 |

Under these conditions, the sampling distribution of the test statistic is $\chi^2$ on 4 degrees of freedom. We will use a chi-square test for independence.

$$\chi^2 = \sum_{all\,cells} \frac{(Obs - Exp)^2}{Exp} \approx 13.849,$$

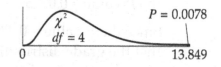

and the *P*-value $\approx 0.0078$

Since the *P*-value $\approx 0.0078$ is low, reject the null hypothesis. There is strong evidence of an association between region and political party in Montana. Residents in the West are more likely to be Democrats than Republicans, and residents in the Northeast are more likely to be Republicans than Democrats.

**43. Grades.**

a) We have two groups, students of Professor Alpha and students of Professor Beta, and we are concerned with the distribution of one variable, grade. We will perform a chi-square test for homogeneity.

b) $H_0$: The distribution of grades is the same for the two professors.

   $H_A$: The distribution of grades is different for the two professors.

c) The expected counts are organized in the table.

Since three cells have expected counts less than 5, the chi-square procedures are not appropriate. Cells would have to be combined in order to proceed. (We will do this in another exercise.)

|  | Prof. Alpha | Prof. Beta |
|---|---|---|
| A | 6.667 | 5.333 |
| B | 12.778 | 10.222 |
| C | 12.222 | 9.778 |
| D | 6.111 | 4.889 |
| F | 2.222 | 1.778 |

**45. Grades again.**

a) **Counted data condition:** The data are counts.

**Randomization condition:** Assume that these students are representative of all students that have ever taken courses from the professors.

**Expected cell frequency condition:** The expected counts are all greater than 5.

|   | Prof. Alpha (Obs/Exp) | Prof. Beta (Obs/Exp) |
|---|---|---|
| **A** | 3 / 6.667 | 9 / 5.333 |
| **B** | 11 / 12.778 | 12 / 10.222 |
| **C** | 14 / 12.222 | 8 / 9.778 |
| **Below C** | 12 / 8.333 | 3 / 6.667 |

b) Under these conditions, the sampling distribution of the test statistic is $\chi^2$ on 3 degrees of freedom, instead of 4 degrees of freedom before the change in the table. We will use a chi-square test for homogeneity.

c) With $\chi^2 = \sum_{all\,cells} \dfrac{(Obs - Exp)^2}{Exp} \approx 9.306$,

the *P*-value $\approx 0.0255$.

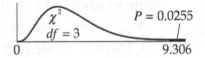

Since the *P*-value = 0.0255 is low, we reject the null hypothesis. There is evidence that the grade distributions for the two professors are different. Professor Alpha gives fewer As and more grades below C than Professor Beta.

**47. Racial steering.**

$H_0$: There is no association between race and section of the complex in which people live.

$H_A$: There is an association between race and section of the complex in which people live.

**Counted data condition:** The data are counts.

**Randomization condition:** Assume that the recently rented apartments are representative of all apartments in the complex.

|   | White (Obs/Exp) | Black (Obs/Exp) |
|---|---|---|
| **Section A** | 87 / 76.179 | 8 / 18.821 |
| **Section B** | 83 / 93.821 | 34 / 23.179 |

**Expected cell frequency condition:** The expected counts are all greater than 5.

Under these conditions, the sampling distribution of the test statistic is $\chi^2$ on 1 degree of freedom. We will use a chi-square test for independence.

$$\chi^2 = \sum_{all\,cells} \frac{(Obs - Exp)^2}{Exp} \approx \frac{(87 - 76.179)^2}{76.179} + \frac{(8 - 18.821)^2}{18.821} + \frac{(83 - 93.821)^2}{93.821} + \frac{(34 - 23.179)^2}{23.179}$$

$$\approx 1.5371 + 6.2215 + 1.2481 + 5.0517$$

$$\approx 14.058$$

With $\chi^2 \approx 14.058$, on 1 degree of freedom, the *P*-value $\approx 0.0002$.

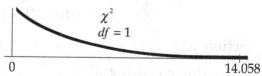

Since the *P*-value $\approx 0.0002$ is low, we reject the null hypothesis. There is strong evidence of an association between race and the section of the apartment complex in which people live. An examination of the components shows us that whites are much more likely to rent in Section A (component = 6.2215), and blacks are much more likely to rent in Section B (component = 5.0517).

### 49. Pregnancies.

$H_0$: Pregnancy outcome is independent of age.

$H_A$: There is an association between pregnancy outcome and age.

**Counted data condition:** The data are counts.
**Randomization condition:** Assume that these women are representative of all pregnant women.
**Expected cell frequency condition:** The expected counts are all greater than 5.

|  | Early Preterm (Obs / Exp) | Late Preterm (Obs / Exp) |
|---|---|---|
| **Under 20** | 129 / 116.55 | 270 / 282.45 |
| **20 – 29** | 243 / 249.75 | 612 / 605.25 |
| **30 – 39** | 165 / 172.05 | 424 / 416.95 |
| **40 and over** | 18 / 16.65 | 39 / 40.35 |

Under these conditions, the sampling distribution of the test statistic is $\chi^2$ on 3 degrees of freedom. We will use a chi-square test for independence.

With $\chi^2 = \sum_{all\,cells} \frac{(Obs - Exp)^2}{Exp} \approx 2.699$, on 3 degrees of freedom, the *P*-value $\approx 0.440$.

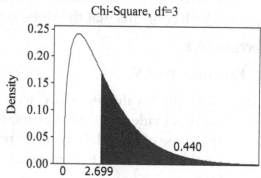

Since the *P*-value is high, we fail to reject the null hypothesis. There is no evidence of an association between pregnancy outcome and age.

## Chapter 20 – Inferences for Regression

**Section 20.1**

**1. Graduate earnings.**

The linear regression model that predicts earnings from SAT score is $\widehat{Earn} = 14468.1 + 27.262(SAT)$. This model predicts that graduates earn, on average, \$27.26 per year more for each additional point in SAT score.

**Section 20.2**

**3. Earnings II.**

The residual plot has no structure, and there are not any striking outliers. The histogram of residuals is symmetric and bell-shaped. All conditions are satisfied to proceed with the regression analysis.

**Section 20.3**

**5. Earnings, part III.**

The error standard deviation is $s = \$5603$. This is the standard deviation of the residuals, which helps us to understand the amount of spread about the regression model.

**7. Earnings, part IV.**

The standard error for the slope tells us how much the slope of the regression equation would vary from sample to sample. If we took many samples, and found a regression equation for each, we would expect most of these slopes to be within a couple of standard deviations of the true slope of the association between graduation rate and acceptance rate. The P-value is essentially zero, which confirms that the slope is statistically significantly different than zero.

**Section 20.4**

**9. Earnings, part V.**

Since the *P*-value is so low, we reject the null hypothesis. We can conclude that there is evidence of a linear relationship between *Earn* and *SAT*. In other words, the slope of the relationship is not zero. It seems that the those who score higher on their SAT tend to earn more.

**11. Earnings, part VI.**

$$b_1 \pm t^*_{n-2} \times SE(b_1) = 27.2642 \pm (t^*_{704}) \times 1.545 \approx (24.23, \ 30.30)$$

We are 95% confident that the true slope relating *Earn* and *SAT* is between 24.23 and 30.30 dollars per year per SAT point. In other words, for each additional point of SAT score, the model predicts an increase in earnings of between 24.23 and 30.30 dollars per year.

## Section 20.5

### 13. Earnings and need.

a) $\widehat{Earn} = 23974.2 + 23.1880(SAT) - 8500.75(\%need)$

b) The model predicts that, on average, *Earn* is expected to increase by approximately $23.18 per year for each additional SAT point, after allowing for the effects of *%need*. This differs from the previous interpretation by taking the variable, *%need* into account.

### 15. Earnings and more.

a) *ACT* and *SAT* are highly correlated with each other. After all, they are very similar measures. Thus *SAT*, after allowing for the effect of *ACT*, is not really a measure of test performance but rather a measure of how students who take the *SAT* may differ from those who take the *ACT* at the colleges in question.

b) $b_1 \pm t^*_{n-(k+1)} \times SE(b_1) = 10.1117 \pm (t^*_{683}) \times 4.336 \approx (1.56, 18.66)$

We are 95% confident that the coefficient of *SAT* in the linear relationship predicting *Earn* from *SAT*, *%need*, and *ACT* is between 1.56 and 18.66. This interval contains much smaller values as plausible values.

c) We are less confident that this coefficient is different than zero. The collinearity with *ACT* has inflated the variance of the coefficient of *SAT*, leading to a smaller t-ratio and larger P-value.

## Section 20.6

### 17. Earnings, predictions.

A prediction interval for an individual SAT score will be too wide to be of much use in predicting future earnings.

## Chapter Exercises.

### 19. Earnings, planning.

No, regression models describe the data as they are. They cannot predict what would happen if a change were made. We cannot conclude that earning a higher SAT score will lead to higher earnings; that would suppose a causal relationship that clearly isn't true.

### 21. Tracking hurricanes 2015.

a) The equation of the line of best fit for these data points is

$\widehat{Error\_24th} = 133.024 - 2.05999(Year)$, where *Year* is measured in years since 1970. According to the linear model, the error made in predicting a hurricane's path was about 133 nautical miles, on average, in 1970. It has been declining at rate of about 2.06 nautical miles per year.

**b)** $H_0$: There has been no change in prediction accuracy.  $(\beta_1 = 0)$

$H_A$: There has been a change in prediction accuracy.  $(\beta_1 \neq 0)$

**c)** Assuming the conditions have been met, the sampling distribution of the regression slope can be modeled by a Student's *t*-model with 46 – 2 = 44 degrees of freedom.  We will use a regression slope *t*-test.

The value of *t* = –11.9.  The *P*-value ≤ 0.0001 means that the association we see in the data is unlikely to occur by chance.  We reject the null hypothesis, and conclude that there is strong evidence that the prediction accuracies have in fact been changing during the time period.

**d)** 76.2% of the variation in the prediction accuracy is accounted for by the linear model based on year.

### 23. Sea ice.

**a)** $\widehat{Extent} = 73.7928 - 4.42138(MeanGlobalTemp)$.  According to the model, we would expect 73.79 square kilometers of sea ice in the northern Artic if the mean global temperature were 0 degrees Celsius. The model predicts a decrease of 4.42 square kilometers of sea ice in the northern Arctic for each additional degree in mean global temperature.

**b)** **Straight enough condition:** The scatterplot shows a moderate, linear relationship with no striking outliers, but the residuals plot shows a possible bend.
**Does the plot thicken? condition:** The residuals plot shows slightly greater variability on the left than on the right.
**Nearly Normal condition:** The Normal probability plot is reasonably straight.

We should proceed with caution because the conditions are almost satisfied.

**c)** The standard deviation of the residuals is 0.68295.

**d)** The standard error of the slope of the regression line is 0.5806.

**e)** The standard error is the estimated standard deviation of the sampling distribution of the slope coefficient. Over many random samples from this population (or with a bootstrap), we'd expect to see slopes of the samples varying by this much.

**f)** No. There is evidence of an association between the mean global temperature and the extent of sea ice in the northern Artic, but we cannot establish causation from this study.

## 25. More sea ice.

Since conditions have been satisfied in a previous exercise, the sampling distribution of the regression slope can be modeled by a Student's *t*-model with $(37 - 2) = 35$ degrees of freedom.

$$b_1 \pm t^*_{n-2} \times SE(b_1) = -4.42138 \pm (t^*_{35}) \times 0.5806 \approx (-5.60, -3.24)$$

We are 95% confident that the extent of sea ice in the northern artic is decreasing at a rate of between 3.24 and 5.60 square kilometers for each additional degree Celsius in mean global temperature.

## 27. Hot dogs.

a)  $H_0$: There's no association between calories and sodium content of all-beef hot dogs. $(\beta_1 = 0)$

 $H_A$: There is an association between calories and sodium content. $(\beta_1 \neq 0)$

b)  Assuming the conditions have been met, the sampling distribution of the regression slope can be modeled by a Student's *t*-model with $(13 - 2) = 11$ degrees of freedom. We will use a regression slope *t*-test. The equation of the line of best fit for these data points is: $\widehat{Sodium} = 90.9783 + 2.29959(Calories)$

The value of $t = 4.10$. The *P*-value of 0.0018 means that the association we see in the data is very unlikely to occur by chance alone. We reject the null hypothesis, and conclude that there is evidence of a linear association between the number of calories in all-beef hotdogs and their sodium content. Because of the positive slope, there is evidence that hot dogs with more calories generally have higher sodium contents.

## 29. Second frank.

a)  Among all-beef hot dogs with the same number of calories, the sodium content varies, with a standard deviation of about 60 mg.

b)  The standard error of the slope of the regression line is 0.5607 milligrams of sodium per calorie.

c)  If we tested many other samples of all-beef hot dogs, the slopes of the resulting regression lines would be expected to vary, with a standard deviation of about 0.56 mg of sodium per calorie.

## 31. Last dog.

$$b_1 \pm t^*_{n-2} \times SE(b_1) = 2.29959 \pm (2.201) \times 0.5607 \approx (1.06, 3.54)$$

We are 95% confident that for every additional calorie, all-beef hot dogs have, on average, between 1.06 and 3.54 mg more sodium.

### 33. Marriage age 2015.

a) $H_0$: The difference in age between men and women at first marriage has not been decreasing since 1975. $(\beta_1 = 0)$

$H_A$: The difference in age between men and women at first marriage has been decreasing since 1975. $(\beta_1 < 0)$

b) **Straight enough condition:** The scatterplot is straight enough.
**Independence assumption:** We are examining a relationship over time, so there is reason to be cautious, but the residuals plot shows no evidence of dependence.
**Does the plot thicken? condition:** The residuals plot shows no obvious trends in the spread.
**Nearly Normal condition, Outlier condition:** The histogram is reasonably unimodal and symmetric, and shows no obvious skewness or outliers. The normal probability plot is somewhat curved, so we should be cautious when making claims about the linear association.

c) Since conditions have been satisfied, the sampling distribution of the regression slope can be modeled by a Student's $t$-model with $37 - 2 = 35$ degrees of freedom. We will use a regression slope $t$-test. The equation of the line of best fit for these data points is: $\widehat{(Men - Women)} = 33.6043 - 0.015821(Year)$

The value of $t = -15.9$. The $P$-value of less than 0.0001 (even though this is the value for a two-tailed test, it is still very small) means that the association we see in the data is unlikely to occur by chance. We reject the null hypothesis, and conclude that there is strong evidence of a negative linear relationship between difference in age at first marriage and year. The difference in marriage age between men and women appears to be decreasing over time.

### 35. Marriage age 2015, again.

$$b_1 \pm t_{n-2}^* \times SE(b_1) = -0.015821 \pm t_{35}^* \times 0.0010 \approx (-0.018, -0.014)$$

We are 95% confident that the mean difference in age between men and women at first marriage decreases by between 0.014 and 0.018 years in age for each year that passes.

### 37. Streams.

a) $H_0$: There is no linear relationship between BCI and pH. $(\beta_1 = 0)$
$H_A$: There is a linear relationship between BCI and pH. $(\beta_1 \neq 0)$

**b)** Assuming the conditions for inference are satisfied, the sampling distribution of the regression slope can be modeled by a Student's $t$-model with $163 - 2 = 161$ degrees of freedom. We will use a regression slope $t$-test. The equation of the line of best fit for these data points is: $\widehat{BCI} = 2733.37 - 197.694(pH)$.

$$t = \frac{b_1 - \beta_1}{SE(b_1)}$$

$$t = \frac{-197.694 - 0}{25.57}$$

$$t \approx -7.73$$

The value of $t \approx -7.73$. The $P$-value (two-sided!) is less than 0.0001

**c)** Since the $P$-value is so low, we reject the null hypothesis, and conclude that there is strong evidence of a linear relationship between BCI and pH. Streams with higher pH tend to have lower BCI.

**39. Streams again.**

**a)** $H_0$: After accounting for Alkali content, *BCI* and *pH* are not (linearly) related. $(\beta_1 = 0)$

$H_A$: After accounting for Alkali content, *BCI* and *pH* are associated. $(\beta_1 \neq 0)$

**b)** With a very large P-value of 0.47, we fail to reject H0. We failed to find evidence that *pH* is related to *BCI* after allowing for the effects of *Alkali*.

**c)** *pH* is likely to be correlated with *Alkali*. After allowing for the effect of *Alkali*, there is no remaining effect of *pH*. The collinearity has inflated the standard error of the coefficient of *pH*, reducing its $t$-ratio.

**d)** The slope of this plot is -1.45740; because a partial regression plot will have the same slope as the corresponding coefficient in the multiple regression model.

**e)** The point on the far left of the plot is influential, so the least squares slope of the partial regression plot is larger (closer to zero) than it would be without it. The coefficient of *Alkali* larger than it would be if the point were not included.

**41. Climate change 2016.**

**a)** $\widehat{Temp} = -3.17933 + 0.00992(CO_2)$

**b)** $H_0$: There is no linear relationship between $CO_2$ level and global temperature. $(\beta_1 = 0)$

$H_A$: There is a linear relationship between $CO_2$ level and global temperature. $(\beta_1 \neq 0)$

**Straight enough condition:** The scatterplot is straight enough to try a linear model.
**Independence assumption:** The residuals plot is scattered.
**Does the plot thicken? condition:** The spread of the residuals is consistent.
**Nearly Normal condition, Outlier condition:** The histogram of residuals is reasonably unimodal and symmetric.

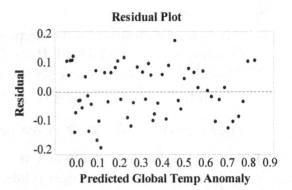

**Residual Plot**

Since conditions have been satisfied, the sampling distribution of the regression slope can be modeled by a Student's $t$-model with $(58 - 2) = 56$ degrees of freedom. We will use a regression slope $t$-test. The equation of the line of best fit for these data points is: $\widehat{Temp} = -3.17933 + 0.00992(CO_2)$.

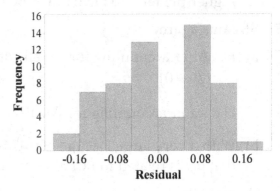

The value of $t = 22.1$. The $P$-value of less than 0.0001 means that the association we see in the data is unlikely to occur by chance. We reject the null hypothesis, and conclude that there is strong evidence of a linear relationship between $CO_2$ level and global temperature. Higher $CO_2$ levels are associated with higher global temperature anomaly.

c) They may be useful. The standard deviation of the residuals is small and the plot is straight. $R^2 = 89.7\%$, so 89.7% of the variability in global temperature anomaly is accounted for by the linear relationship.

d) No, the is does not prove that increasing $CO_2$ levels are causing global warming. The model is consistent with the claim that increasing $CO_2$ is causing global climate change, but it does not by itself prove that this is the mechanism. Other scientific studies showing how $CO_2$ can trap heat are necessary for that.

**43. Climate change, part II.**

a) Since conditions have been satisfied previously, the sampling distribution of the regression slope can be modeled by a Student's $t$-model with $(58 - 2) = 56$ degrees of freedom. We will use a regression slope $t$-interval, with 90% confidence.

$$b_1 \pm t^*_{n-2} \times SE(b_1) = 0.00992 \pm t^*_{58} \times 0.0004 \approx (0.0092, 0.011)$$

**b)** We are 90% confident that the mean global temperature anomaly increases by between 0.0092 and 0.011 degrees per ppm of $CO_2$.

## 45. Climate change, again.

**a)** The regression equation predicts that a $CO_2$ level of 450 ppm will have a mean global temperature anomaly of $-3.17933 + 0.00992(450) = 1.28467$ degrees Celsius.

$$\hat{y}_v \pm t^*_{n-2}\sqrt{SE^2(b_1)\cdot(x_v-\bar{x})^2 + \frac{s_e^2}{n}}$$

$$= 1.28467 \pm t^*_{56}\sqrt{0.0004^2 \cdot (450-352.566)^2 + \frac{0.0885^2}{58}} \approx (1.19,\ 1.37)$$

We are 90% confident that years in which $CO_2$ levels are 450 ppm will have a mean global temperature anomaly of between 1.19 and 1.37 degrees Celsius.

(This interval was calculated from the original data set, using technology. Minor differences due to level of precision in various inputs are fine. These differences are unlikely to change our conclusions about the association.)

**b)** The regression equation predicts that a $CO_2$ level of 450 ppm will have a mean global temperature anomaly of $-3.17933 + 0.00992(450) = 1.28467$ degrees Celsius.

$$\hat{y}_v \pm t^*_{n-2}\sqrt{SE^2(b_1)\cdot(x_v-\bar{x})^2 + \frac{s_e^2}{n} + s_e^2}$$

$$= 1.28467 \pm t^*_{56}\sqrt{0.0004^2 \cdot (450-352.566)^2 + \frac{0.0885^2}{58} + 0.0885^2} \approx (1.08,\ 1.48)$$

We are 90%confident that a year in which $CO_2$ levels are 450 ppm will have a global temperature anomaly of between 1.08 and 1.48 degrees Celsius.

(This interval was calculated from the original data set, using technology. Minor differences due to level of precision in various inputs are fine. These differences are unlikely to change our conclusions about the association.)

**c)** Yes, 1.3 degrees Celsius is a plausible value, since it is within the interval.

## 47. Brain size.

**a)** $H_0$: There is no linear relationship between brain size and IQ. $(\beta_1 = 0)$

$H_A$: There is a linear relationship between brain size and IQ. $(\beta_1 \neq 0)$

Since these data were judged acceptable for inference, the sampling distribution of the regression slope can be modeled by a Student's $t$-model with $(21 - 2) = 19$ degrees of freedom. We will use a regression slope $t$-test. The equation of the line of best fit for these data points is: $\widehat{IQ\_Verbal} = 24.1835 + 0.0988(Size)$.

The value of $t \approx$ 1.12. The $P$-value of 0.2775 means that the association we see in the data is likely to occur by chance. We fail to reject the null hypothesis, and conclude that there is no evidence of a linear relationship between brain size and verbal IQ score.

$$t = \frac{b_1 - \beta_1}{SE(b_1)}$$

$$t = \frac{0.0988 - 0}{0.0884}$$

$$t \approx 1.12$$

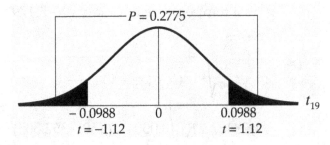

b) Since $R^2 = 6.5\%$, only 6.5% of the variability in verbal IQ can be accounted for by brain size. This association is very weak. There are three students with large brains who scored high on the IQ test. Without them, there appears to be no association at all.

## 49. City climate.

**Straight enough condition:** The scatterplot is not straight.
**Independence assumption:** The residuals plot shows a curved pattern.
**Does the plot thicken? condition:** The spread of the residuals is not consistent. The residuals plot shows decreasing variability as the predicted values increase.
**Nearly Normal condition, Outlier condition:** The histogram of residuals is skewed to the right, with an outlier.

These data are not appropriate for inference.

## 51. Ozone and population.

a) $H_0$: There is no linear relationship between population and ozone level. $(\beta_1 = 0)$

$H_A$: There is a linear relationship between population and ozone level. $(\beta_1 \neq 0)$

Assuming the conditions for inference are satisfied, the sampling distribution of the regression slope can be modeled by a Student's $t$-model with $16 - 2 = 14$ degrees of freedom. We will use a regression slope $t$-test. The equation of the line of best fit for these data points is: $\widehat{Ozone} = 18.892 + 6.650(Population)$, where ozone level is measured in parts per million and population is measured in millions.

The value of $t \approx 3.48$.

The *P*-value of 0.0037 means that the association we see in the data is unlikely to occur by chance. We reject the null hypothesis, and conclude that there is strong evidence of a positive linear relationship between ozone level and population. Cities with larger populations tend to have higher ozone levels.

$$t = \frac{b_1 - \beta_1}{SE(b_1)}$$

$$t = \frac{6.650 - 0}{1.910}$$

$$t \approx 3.48$$

**b)** City population is a good predictor of ozone level. Population explains 84% of the variability in ozone level and *s* is just over 5 parts per million.

## 53. Ozone, again

**a)** $b_1 \pm t^*_{n-2} \times SE(b_1) = 6.65 \pm (1.761) \times 1.910 \approx (3.29, 10.01)$

We are 90% confident that each additional million people will increase mean ozone levels by between 3.29 and 10.01 parts per million.

**b)** The regression equation predicts that cities with a population of 600,000 people will have ozone levels of $18.892 + 6.650(0.6) = 22.882$ parts per million.

$$\hat{y}_v \pm t^*_{n-2} \sqrt{SE^2(b_1) \cdot (x_v - \bar{x})^2 + \frac{s_e^2}{n}}$$

$$= 22.882 \pm (1.761) \sqrt{1.91^2 \cdot (0.6 - 1.7)^2 + \frac{5.454^2}{16}} \approx (18.47, 27.29)$$

We are 90% confident that the mean ozone level for cities with populations of 600,000 will be between 18.47 and 27.29 parts per million.

## 55. Tablet computers 2014.

**a)** Since there are $34 - 2 = 32$ degrees of freedom, there were 34 tablet computers tested.

**b)** **Straight enough condition:** The scatterplot is roughly straight, but scattered.
**Independence assumption:** The residuals plot shows no pattern.
**Does the plot thicken? condition:** The spread of the residuals is consistent.
**Nearly Normal condition:** The Normal probability plot of residuals is reasonably straight.

**c)** H₀: There is no linear relationship between maximum brightness and battery life. $(\beta_1 = 0)$

Hₐ: There is a positive linear relationship between maximum brightness and battery life. $(\beta_1 > 0)$

Since the conditions for inference are satisfied, the sampling distribution of the regression slope can be modeled by a Student's $t$-model with $34 - 2 = 32$ degrees of freedom. We will use a regression slope $t$-test. The equation of the line of best fit for these data points is: $\widehat{Hours} = 5.38719 + 0.00904(ScreenBrightness)$, battery life measure in hours and screen brightness measured in $cd/m^2$.

The value of $t \approx 2.02$. The P-value of 0.0522 means that the association we see in the data is not unlikely to occur by chance. We fail to reject the null hypothesis, and conclude that there is little evidence of a positive linear relationship between battery life and screen brightness.

d) Since $R^2 = 11.3\%$, only 11.3% of the variability in battery life can be accounted for by screen brightness. The residual standard deviation is 2.128 hours. That's pretty large, considering the range of battery life is only about 9 hours. Even if we concluded that there was some evidence of a linear association, it is too weak to be of much use. Predictions would tend to be very imprecise.

e) The equation of the line of best fit for these data points is:
$\widehat{Hours} = 5.38719 + 0.00904(ScreenBrightness)$, battery life measure in hours and screen brightness measured in $cd/m^2$.

f) There are 32 degrees of freedom, so use $t^*_{32} = 1.694$.

$$b_1 \pm t^*_{n-2} \times SE(b_1) = 0.00904 \pm (1.694) \times 0.0045 \approx (0.00142, 0.0167)$$

g) We are 90% confident that the mean battery life increases by between 0.00142 and 0.0167 hours for each additional $cd/m^2$ of screen brightness.

## 57. Midterms.

a) The regression output is to the right. The model is:

$$\widehat{Midterm2} = 12.005 + 0.721(Midterm1)$$

Dependent variable is: Midterm 2
No Selector
R squared = 19.9%   R squared (adjusted) = 18.6%
s = 16.78 with 64 - 2 = 62 degrees of freedom

| Source | Sum of Squares | df | Mean Square | F-ratio |
|---|---|---|---|---|
| Regression | 4337.14 | 1 | 4337.14 | 15.4 |
| Residual | 17459.5 | 62 | 281.604 | |

| Variable | Coefficient | s.e. of Coeff | t-ratio | prob |
|---|---|---|---|---|
| Constant | 12.0054 | 15.96 | 0.752 | 0.4546 |
| Midterm 1 | 0.720990 | 0.1837 | 3.92 | 0.0002 |

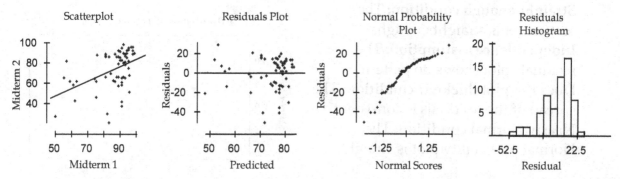

b) **Straight enough condition:** The scatterplot shows a weak, positive relationship between Midterm 2 score and Midterm 1 score. There are several outliers, but removing them only makes the relationship slightly stronger. The relationship is straight enough to try linear regression.

**Independence assumption:** The residuals plot shows no pattern..

**Does the plot thicken? condition:** The spread of the residuals is consistent.

**Nearly Normal condition, Outlier condition:** The histogram of the residuals is unimodal, slightly skewed with several possible outliers. The Normal probability plot shows some slight curvature.

Since we had some difficulty with the conditions for inference, we should be cautious in making conclusions from these data. The small $P$-value of 0.0002 for the slope would indicate that the slope is statistically distinguishable from zero, but the $R^2$ value of 0.199 suggests that the relationship is weak. Midterm 1 isn't a useful predictor of Midterm 2.

c) The student's reasoning is not valid. The $R^2$ value is only 0.199 and the value of $s$ is 16.8 points. Although correlation between Midterm 1 and Midterm 2 may be statistically significant, it isn't of much practical use in predicting Midterm 2 scores. It's too weak.

**59. Strike two.**

$H_0$: Effectiveness is independent of the player's initial ability. $(\beta_1 = 0)$

$H_A$: Effectiveness of the video depends on the player's initial ability. $(\beta_1 \neq 0)$

**Straight enough condition:** The scatterplot is straight enough.
**Independence assumption:** The residuals plot shows no pattern.
**Does the plot thicken? condition:** The spread of the residuals is consistent.
**Nearly Normal condition:** The Normal probability plot is straight.

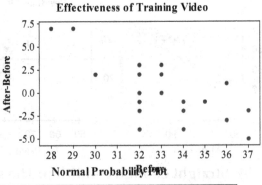

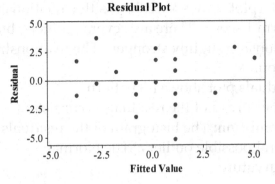

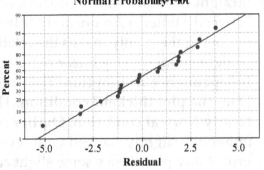

Since the conditions for inference are inference are satisfied, the sampling distribution of the regression slope can be modeled by a Student's *t*-model with 20 – 2 = 18 degrees of freedom. We will use a regression slope *t*-test.

The equation of the line of best fit for these data points is:

$\overline{(After - Before)} = 32.316 - 0.9748(Before)$, where we are counting the number of strikes thrown before and after the training program.

The value of $t \approx -4.34$. Since the *P*-value is 0.004, reject the null hypothesis, and conclude that there is evidence of a linear relationship between the player's intial ability and the effectiveness of the program. The negative slope indicates that the method is more effective for those whose initial performance was poorest and less effective for those whose initial performance was better. This appears to be a case of regression to the mean. Those who were above average initially tended to do worse after training. Those who were below average initially tended to improve.

**61. Education and mortality.**

**a)** **Straight enough condition:** The scatterplot is straight enough.
**Independence assumption:** The residuals plot shows no pattern. If these cities are representative of other cities, we can generalize our results.
**Does the plot thicken? condition:** The spread of the residuals is consistent.
**Nearly Normal condition, Outlier condition:** The histogram of the residuals is unimodal and symmetric with no outliers.

b) $H_0$: There is no linear relationship between education and mortality. $(\beta_1 = 0)$

   $H_A$: There is a linear relationship between education and mortality. $(\beta_1 \neq 0)$

   Since the conditions for inference are inference are satisfied, the sampling distribution of the regression slope can be modeled by a Student's *t*-model with 58 – 2 = 56 degrees of freedom. We will use a regression slope *t*-test. The equation of the line of best fit for these data points is:

   $\overline{Mortality} = 1493.26 - 49.9202(Education)$.

   The value of $t \approx -6.24$. The *P*-value of essentially 0 means that the association we see in the data is unlikely to occur by chance. We reject the null hypothesis, and conclude that there is strong evidence of a linear relationship between the level of education in a city and its mortality rate. Cities with lower education levels tend to have higher mortality rates.

   $$t = \frac{b_1 - \beta_1}{SE(b_1)}$$

   $$t = \frac{-49.9202 - 0}{8.000}$$

   $$t \approx -6.24$$

c) We cannot conclude that getting more education is likely to prolong your life. Association does not imply causation. There may be lurking variables involved.

d) For 95% confidence, $t_{56}^* \approx 2.00327$.

   $$b_1 \pm t_{n-2}^* \times SE(b_1) = -49.9202 \pm (2.003) \times 8.000 \approx (-65.95, -33.89)$$

e) We are 95% confident that the mean number of deaths per 100,000 people decreases by between 33.89 and 65.95 deaths for an increase of one year in average education level.

f) The regression equation predicts that cities with an adult population with an average of 12 years of school will have a mortality rate of $1493.26 - 49.9202(12) = 894.2176$ deaths per 100,000. The average education level was 11.0328 years.

   $$\hat{y}_v \pm t_{n-2}^* \sqrt{SE^2(b_1) \cdot (x_v - \bar{x})^2 + \frac{s_e^2}{n}}$$

   $$= 894.2176 \pm (2.003) \sqrt{8.00^2 \cdot (12 - 11.0328)^2 + \frac{47.92^2}{58}} \approx (874.239, \ 914.196)$$

   We are 95% confident that the mean mortality rate for cities with an average of 12 years of schooling is between 874.239 and 914.196 deaths per 100,000 residents.

## 63. Right-to-work laws.

a) $\widehat{\text{Logit}(Embrace)} = 0.5796 - 0.0149age$

b) Yes, the P-value is < 0.01, which means there is strong evidence to suggest the association.

c) The coefficient on age is negative, so an older person is less likely to respond "Embrace."

# Review of Part V – Inference for Relationships

### R5.1. Herbal cancer.

$H_0$: The cancer rate for those taking the herb is the same as the cancer rate for those not taking the herb. $\left(p_{Herb} = p_{Not} \text{ or } p_{Herb} - p_{Not} = 0\right)$

$H_A$: The cancer rate for those taking the herb is higher than the cancer rate for those not taking the herb. $\left(p_{Herb} > p_{Not} \text{ or } p_{Herb} - p_{Not} > 0\right)$

### R5.3. Surgery and germs.

a) Lister imposed a treatment, the use of carbolic acid as a disinfectant. This is an experiment.

b) $H_0$: The survival rate when carbolic acid is used is the same as the survival rate when carbolic acid is not used. $\left(p_C = p_N \text{ or } p_C - p_N = 0\right)$

$H_A$: The survival rate when carbolic acid is used is greater than the survival rate when carbolic acid is not used. $\left(p_C > p_N \text{ or } p_C - p_N > 0\right)$

**Randomization condition:** There is no mention of random assignment. Assume that the two groups of patients were similar, and amputations took place under similar conditions, with the use of carbolic acid being the only variable.
**Independent samples condition:** It is reasonable to think that the groups were not related in any way.
**Success/Failure condition:** $n\hat{p}$ (carbolic acid) = 34, $n\hat{q}$ (carbolic acid) = 6, $n\hat{p}$ (none) = 19, and $n\hat{q}$ (none) = 16. The number of patients who died in the carbolic acid group is only 6, but the expected number of deaths using the pooled proportion, $n\hat{q}_{pooled} = (40)(\frac{22}{75}) = 11.7$, so the samples are both large enough.

Since the conditions have been satisfied, we will perform a two-proportion $z$-test. We will model the sampling distribution of the difference in proportion with a Normal model with mean 0 and standard deviation estimated by:

$$SE\left(\hat{p}_C - \hat{p}_N\right) = \sqrt{\frac{\hat{p}_C \hat{q}_C}{n_C} + \frac{\hat{p}_N \hat{q}_N}{n_N}} = \sqrt{\frac{\left(\frac{34}{40}\right)\left(\frac{6}{40}\right)}{40} + \frac{\left(\frac{19}{35}\right)\left(\frac{16}{35}\right)}{35}} \approx 0.10137987.$$

The observed difference between the proportions is:
$\frac{34}{40} - \frac{19}{35} = 0.3071429.$

Since the P-value = 0.0012 is low, we reject the null hypothesis. There is strong evidence that the survival rate is higher when carbolic acid is used to disinfect the operating room than when carbolic acid is not used.

$$z \approx \frac{0.3071429 - 0}{0.10137987}$$

$$t \approx 3.03$$

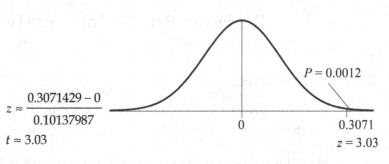

$P = 0.0012$

$0$            $0.3071$

$z = 3.03$

c) We don't know whether or not patients were randomly assigned to treatments, and we don't know whether or not blinding was used.

**R5.5. Twins.**

$H_0$: The proportion of preterm twin births in 2000 is the same as the proportion of preterm twin births in 2010. $\left( p_{2000} = p_{2010} \text{ or } p_{2000} - p_{2010} = 0 \right)$

$H_A$: The proportion of preterm twin births in 1990 is the less than the proportion of preterm twin births in 2000. $\left( p_{2000} < p_{2010} \text{ or } p_{2000} - p_{2010} < 0 \right)$

**Randomization condition:** Assume that these births are representative of all twin births.

**Independent samples condition:** The samples are from different years, so they are unlikely to be related.

**Success/Failure condition:** $n\hat{p}\,(2000) = 20$, $n\hat{q}\,(2000) = 23$, $n\hat{p}\,(2010) = 26$, and $n\hat{q}\,(2010) = 22$ are all greater than 10, so both samples are large enough.

Since the conditions have been satisfied, we will perform a two-proportion z-test. We will model the sampling distribution of the difference in proportion with a Normal model with mean 0 and standard deviation estimated by:

$$SE_{pooled}\left( \hat{p}_{2000} - \hat{p}_{2010} \right) = \sqrt{ \frac{\hat{p}_{pooled}\hat{q}_{pooled}}{n_{2000}} + \frac{\hat{p}_{pooled}\hat{q}_{pooled}}{n_{2010}} } = \sqrt{ \frac{\left(\frac{46}{91}\right)\left(\frac{45}{91}\right)}{43} + \frac{\left(\frac{46}{91}\right)\left(\frac{45}{91}\right)}{48} } \approx 0.1050$$

(You could have chosen to use the unpooled standard error as well. It won't affect your conclusion.)

The observed difference between the proportions is: $0.4651 - 0.5417 = -0.0766$

Since the *P*-value = 0.2329 is high, we fail to reject the null hypothesis. There is no evidence of an increase in the proportion of preterm twin births from 2000 to 2010, at least not at this large city hospital.

$$z = \frac{-0.0766 - 0}{0.1050}$$

$$z \approx -0.73$$

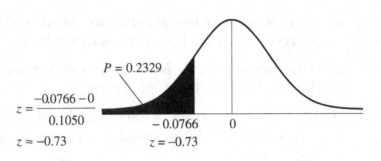

$P = 0.2329$

$-0.0766 \qquad 0$

$z = -0.73$

**R5.7. Eclampsia deaths.**

a) $H_0$: The proportion of pregnant women who die after developing eclampsia is the same for women taking magnesium sulfide as it is for women not taking magnesium sulfide. $(p_{MS} = p_N \;\; \text{or} \;\; p_{MS} - p_N = 0)$

$H_A$: The proportion of pregnant women who die after developing eclampsia is lower for women taking magnesium sulfide than for women not taking magnesium sulfide. $(p_{MS} < p_N \;\; \text{or} \;\; p_{MS} - p_N < 0)$

b) **Randomization condition:** Although not specifically stated, these results are from a large-scale experiment, which was undoubtedly properly randomized. **Independent samples condition:** Subjects were randomly assigned to the treatments.
**Success/Failure condition:** $n\hat{p}$ (mag. sulf.) = 11, $n\hat{q}$ (mag. sulf.) = 29, $n\hat{p}$ (placebo) = 20, and $n\hat{q}$ (placebo) = 76 are all greater than 10, so both samples are large enough.

Since the conditions have been satisfied, we will perform a two-proportion z-test. We will model the sampling distribution of the difference in proportion with a Normal model with mean 0 and standard deviation estimated by:

$$SE_{pooled}\left(\hat{p}_{MS} - \hat{p}_N\right) = \sqrt{\frac{\hat{p}_{pooled}\hat{q}_{pooled}}{n_{MS}} + \frac{\hat{p}_{pooled}\hat{q}_{pooled}}{n_N}} = \sqrt{\frac{\left(\frac{31}{136}\right)\left(\frac{105}{136}\right)}{40} + \frac{\left(\frac{31}{136}\right)\left(\frac{105}{136}\right)}{96}} \approx 0.07895.$$

(We could also use the unpooled standard error.)

c) The observed difference between the proportions is: 0.275 – 0.2083 = 0.0667

Since the *P*-value = 0.8008 is high, we fail to reject the null hypothesis. There is no evidence that the proportion of women who may die after developing eclampsia is lower for women taking magnesium sulfide than for women who are not taking the drug.

$$z = \frac{0.0667 - 0}{0.07895}$$

$$z \approx 0.84$$

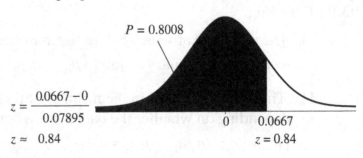

$P = 0.8008$

$0 \qquad 0.0667$

$z = 0.84$

d) There is not sufficient evidence to conclude that magnesium sulfide is effective in preventing death when eclampsia develops.

e) If magnesium sulfide is effective in preventing death when eclampsia develops, then we have made a Type II error.

f) To increase the power of the test to detect a decrease in death rate due to magnesium sulfide, we could increase the sample size or increase the level of significance.

g) Increasing the sample size lowers variation in the sampling distribution, but may be costly. The sample size is already quite large. Increasing the level of significance increases power by increasing the likelihood of rejecting the null hypothesis, but increases the chance of making a Type I error, namely thinking that magnesium sulfide is effective when it is not.

**R5.9. More errors.**

a) Since a treatment (the additive) is imposed, this is an experiment.

b) The company is only interested in an increase in fuel economy, so they will perform a one-sided test.

c) The company will make a Type I error if they decide that the additive increases the fuel economy, when it actually makes no difference in the fuel economy.

d) The company will make a Type II error if they decide that the additive does not increase the fuel economy, when it actually increases fuel economy.

e) The additive manufacturer would consider a Type II error more serious. If the test caused the company to conclude that the manufacturer's product didn't work, and it actually did, the manufacturer would lose sales, and the reputation of their product would suffer.

f) Since this was a controlled experiment, the company can conclude that the additive is the reason that the fuel economy has increased. They should be cautious recommending it for all cars. There is evidence that the additive works well for fleet vehicles, which get heavy use. It might not be effective in cars with a different pattern of use than fleet vehicles.

**R5.11. Crawling.**

a) $H_0$: The mean age at which babies begin to crawl is the same whether the babies were born in January or July. $\left(\mu_{Jan} = \mu_{July} \text{ or } \mu_{Jan} - \mu_{July} = 0\right)$

   $H_A$: There is a difference in the mean age at which babies begin to crawl, depending on whether the babies were born in January or July.
   $\left(\mu_{Jan} \neq \mu_{July} \text{ or } \mu_{Jan} - \mu_{July} \neq 0\right)$

**Independent groups assumption:** The groups of January and July babies are independent.

**Randomization condition:** Although not specifically stated, we will assume that the babies are representative of all babies.

**Nearly Normal condition:** We don't have the actual data, so we can't check the distribution of the sample. However, the samples are fairly large. The Central Limit Theorem allows us to proceed.

Since the conditions are satisfied, it is appropriate to model the sampling distribution of the difference in means with a Student's *t*-model, with 43.68 degrees of freedom (from the approximation formula).

We will perform a two-sample *t*-test. The sampling distribution model has mean 0, with standard error: $SE(\overline{y}_{Jan} - \overline{y}_{July}) = \sqrt{\dfrac{7.08^2}{32} + \dfrac{6.91^2}{21}} \approx 1.9596$.

The observed difference between the mean ages is 29.84 – 33.64 = – 3.8 weeks.

Since the *P*-value = 0.0590 is fairly low, we reject the null hypothesis. There is some evidence that mean age at which babies crawl is different for January and July babies. July babies appear to

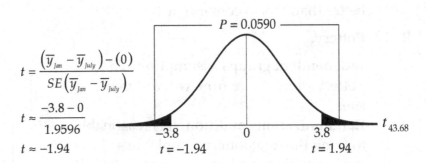

$$t = \frac{\left(\overline{y}_{Jan} - \overline{y}_{July}\right) - (0)}{SE\left(\overline{y}_{Jan} - \overline{y}_{July}\right)}$$

$$t \approx \frac{-3.8 - 0}{1.9596}$$

$$t \approx -1.94$$

crawl a bit earlier than July babies, on average. Since the evidence is not strong, we might want to do some more research into this claim.

**b)** H₀: The mean age at which babies begin to crawl is the same whether the babies were born in April or October. $\left(\mu_{Apr} = \mu_{Oct} \text{ or } \mu_{Apr} - \mu_{Oct} = 0\right)$

H_A: There is a difference in the mean age at which babies begin to crawl, depending on whether the babies were born in April or October. $\left(\mu_{Apr} \neq \mu_{Oct} \text{ or } \mu_{Apr} - \mu_{Oct} \neq 0\right)$

The conditions (with minor variations) were checked in part a.

Since the conditions are satisfied, it is appropriate to model the sampling distribution of the difference in means with a Student's *t*-model, with 59.40 degrees of freedom (from the approximation formula).

We will perform a two-sample *t*-test. The sampling distribution model has mean 0, with standard error: $SE(\bar{y}_{Apr} - \bar{y}_{Oct}) = \sqrt{\dfrac{6.21^2}{26} + \dfrac{7.29^2}{44}} \approx 1.6404$.

The observed difference between the mean ages is 31.84 – 33.35 = – 1.51 weeks.

Since the *P*-value = 0.3610 is high, we fail to reject the null hypothesis. There is no evidence that mean age at which babies crawl is different for April and October babies.

$$t = \frac{(\bar{y}_{Apr} - \bar{y}_{Oct}) - (0)}{SE(\bar{y}_{Apr} - \bar{y}_{Oct})}$$

$$t \approx \frac{-1.51 - 0}{1.6404}$$

$$t \approx -0.92$$

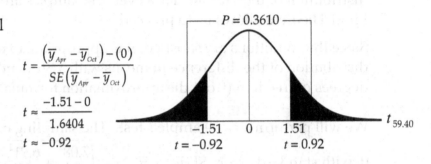

c) These results are not consistent with the researcher's claim. We have slight evidence in one test and no evidence in the other. The researcher will have to do better than this to convince us!

**R5.13. Pottery.**

**Independent groups assumption:** The pottery samples are from two different sites.
**Randomization condition:** It is reasonable to think that the pottery samples are representative of all pottery at that site with respect to aluminum oxide content.
**Nearly Normal condition:** The histograms of aluminum oxide content are roughly unimodal and symmetric.

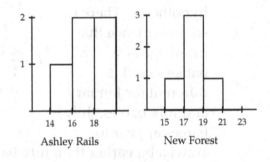

Since the conditions are satisfied, it is appropriate to model the sampling distribution of the difference in means with a Student's *t*-model, with 7 degrees of freedom (from the approximation formula). We will construct a two-sample *t*-interval, with 95% confidence.

$$(\bar{y}_{AR} - \bar{y}_{NF}) \pm t^*_{df} \sqrt{\frac{s^2_{AR}}{n_{AR}} + \frac{s^2_{NF}}{n_{NF}}}$$

$$= (17.32 - 18.18) \pm t^*_7 \sqrt{\frac{1.65892^2}{5} + \frac{1.77539^2}{5}} \approx (-3.37, 1.65)$$

We are 95% confident that the difference in the mean percentage of aluminum oxide content of the pottery at the two sites is between –3.37% and 1.65%. Since 0 is in the interval, there is no evidence that the aluminum oxide content at the two sites is different. It would be reasonable for the archaeologists to think that the same ancient people inhabited the sites.

### R5.15. Feeding fish.

a) If there is no difference in the average fish sizes, the chance of observing a difference this large, or larger, just by natural sampling variation is 0.1%.

b) There is evidence that largemouth bass that are fed a natural diet are larger. The researchers would advise people who raise largemouth bass to feed them a natural diet.

c) If the advice is incorrect, the researchers have committed a Type I error.

### R5.17. Age.

a) **Independent groups assumption:** The group of patients with and without cardiac disease are not related in any way.
**Randomization condition:** Assume that these patients are representative of all people.
**Normal population assumption:** We don't have the actual data, so we will assume that the population of ages of patients is Normal.

Since the conditions are satisfied, it is appropriate to model the sampling distribution of the difference in means with a Student's $t$-model, with 670 degrees of freedom (from the approximation formula). We will construct a two-sample $t$-interval, with 95% confidence.

$$\left(\bar{y}_{Card} - \bar{y}_{None}\right) \pm t^*_{df} \sqrt{\frac{s^2_{Card}}{n_{Card}} + \frac{s^2_{None}}{n_{None}}} = (74.0 - 69.8) \pm t^*_{670} \sqrt{\frac{7.9^2}{450} + \frac{8.7^2}{2397}} \approx (3.39,\ 5.01)$$

We are 95% confident that the mean age of patients with cardiac disease is between 3.39 and 5.01 years higher than the mean age of patients without cardiac disease.

b) Older patients are at greater risk for a variety of health problems. If an older patient does not survive a heart attack, the researchers will not know to what extent depression was involved, because there will be a variety of other possible variables influencing the death rate. Additionally, older patients may be more (or less) likely to be depressed than younger ones.

**R5.19. Eating disorders.**

a) **Randomization condition:** Hopefully, the students were selected randomly.
   **10% condition:** 150 and 200 are less than 10% of all students.
   **Independent samples condition:** The groups are independent.
   **Success/Failure condition:** $n\hat{p}$ (Muslim) = 46, $n\hat{q}$ (Muslim) = 104, $n\hat{p}$ (Christian) = 34, and $n\hat{q}$ (Christian) = 166 are all greater than 10, so the samples are both large enough.

   Since the conditions have been satisfied, we will find a two-proportion $z$-interval.

$$\left(\hat{p}_M - \hat{p}_C\right) \pm z^* \sqrt{\frac{\hat{p}_M \hat{q}_M}{n_M} + \frac{\hat{p}_C \hat{q}_C}{n_C}}$$

$$= \left(\frac{46}{150} - \frac{34}{200}\right) \pm 1.960 \sqrt{\frac{\left(\frac{46}{150}\right)\left(\frac{104}{150}\right)}{150} + \frac{\left(\frac{34}{200}\right)\left(\frac{166}{200}\right)}{200}} = \left(0.046, 0.227\right)$$

   We are 95% confident that the percentage of Muslim students who have an eating disorder is between 4.6 and 22.7 percentage points higher than the percentage of Christian students who have an eating disorder.

b) Although caution in generalizing must be used since the study was restricted to the Spanish city of Ceuta, it appears there is a true difference in the prevalence of eating disorders. We can conclude this because the entire interval is above 0.

**R5.21. Teach for America.**

   $H_0$: The mean score of students with certified teachers is the same as the mean score of students with uncertified teachers. $\left(\mu_C = \mu_U \text{ or } \mu_C - \mu_U = 0\right)$

   $H_A$: The mean score of students with certified teachers is greater than as the mean score of students with uncertified teachers. $\left(\mu_C > \mu_U \text{ or } \mu_C - \mu_U > 0\right)$

   **Independent groups assumption:** The certified and uncertified teachers are independent groups.
   **Randomization condition:** Assume the students studied were representative of all students.
   **Nearly Normal condition:** We don't have the actual data, so we can't look at the graphical displays, but the sample sizes are large, so we can proceed.

   Since the conditions are satisfied, it is appropriate to model the sampling distribution of the difference in means with a Student's $t$-model, with 86 degrees of freedom (from the approximation formula).

   We will perform a two-sample $t$-test. The sampling distribution model has mean 0, with standard error: $SE(\bar{y}_C - \bar{y}_U) = \sqrt{\dfrac{9.31^2}{44} + \dfrac{9.43^2}{44}} \approx 1.9977$.

   The observed difference between the mean scores is $35.62 - 32.48 = 3.14$.

Since the P-value = 0.0598 is fairly low, we reject the null hypothesis. There is some evidence that students with certified teachers had mean scores higher than

$$t = \frac{(\bar{y}_c - \bar{y}_U) - (0)}{SE(\bar{y}_c - \bar{y}_U)}$$

$$t \approx \frac{3.14}{1.9977}$$

$$t \approx 1.57$$

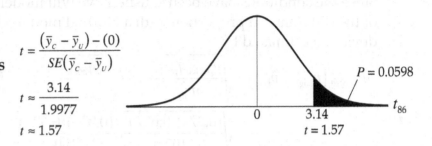

students with uncertified teachers. However, since the P-value is not extremely low, further investigation is recommended.

**R5.23. Teach for America, Part II.**

$H_0$: The mean score of students with certified teachers is the same as the mean score of students with uncertified teachers. $\left(\mu_C = \mu_U \text{ or } \mu_C - \mu_U = 0\right)$

$H_A$: The mean score of students with certified teachers is different than the mean score of students with uncertified teachers. $\left(\mu_C \neq \mu_U \text{ or } \mu_C - \mu_U \neq 0\right)$

**Mathematics:** Since the P-value = 0.002 is low, we reject the null hypothesis. There is strong evidence that students with certified teachers have different mean math scores than students with uncertified teachers. Students with certified teachers do better.

**Language:** Since the P-value = 0.045 is fairly low, we reject the null hypothesis. There is evidence that students with certified teachers have different mean language scores than students with uncertified teachers. Students with certified teachers do better. However, since the P-value is not extremely low, further investigation is recommended.

**R5.25. Insulin and diet.**

a) $H_0$: People with high dairy consumption have IRS at the same rate as those with low dairy consumption. $\left(p_{High} = p_{Low} \text{ or } p_{High} - p_{Low} = 0\right)$

$H_A$: People with high dairy consumption have IRS at a different rate than those with low dairy consumption. $\left(p_{High} \neq p_{Low} \text{ or } p_{High} - p_{Low} \neq 0\right)$

**Random condition:** Assume the people studied are representative of all people.
**10% condition:** 102 and 190 are both less than 10% of all people.
**Independent samples condition:** The two groups are not related.
**Success/Failure condition:** $n\hat{p}$ (high) = 24, $n\hat{q}$ (high) = 78, $n\hat{p}$ (low) = 85, and $n\hat{q}$ (low) = 105 are all greater than 10, so the samples are both large enough.

Since the conditions have been satisfied, we will model the sampling distribution of the difference in proportion with a Normal model with mean 0 and standard deviation estimated by

$$SE_{pooled}\left(\hat{p}_{High} - \hat{p}_{Low}\right) = \sqrt{\frac{\hat{p}_{pooled}\hat{q}_{pooled}}{n_{High}} + \frac{\hat{p}_{pooled}\hat{q}_{pooled}}{n_{Low}}}$$

$$= \sqrt{\frac{(0.373)(0.627)}{102} + \frac{(0.373)(0.627)}{190}} \approx 0.05936$$

(We could also use the unpooled standard error.)

The observed difference between the proportions is $0.2352 - 0.4474 = -0.2122$.

Since the P-value = 0.0004 is very low, we reject the null hypothesis. There is strong evidence that the proportion of people with IRS is different for those who with high dairy consumption compared to those with low dairy

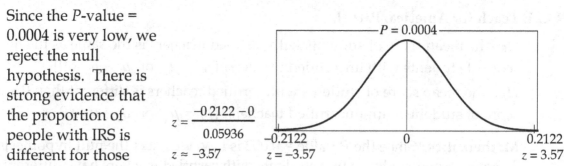

$$z = \frac{-0.2122 - 0}{0.05936}$$

$$z \approx -3.57$$

consumption. People who consume dairy products more than 35 times per week appear less likely to have IRS than those who consume dairy products fewer than 10 times per week.

**b)** There is evidence of an association between the low consumption of dairy products and IRS, but that does not prove that dairy consumption influences the development of IRS. This is an observational study, and a controlled experiment is required to prove cause and effect.

**R5.27. Cloud seeding re-expressed.**

**a)** Histograms of the cloud seeding data appear below. Both distributions are skewed to the right. Most clouds produced a moderate amount of rain, but some clouds, unseeded and seeded alike, produced much large amounts of rain. The two-sample *t*-interval constructed above is not appropriate, since the Nearly Normal Condition is not met.

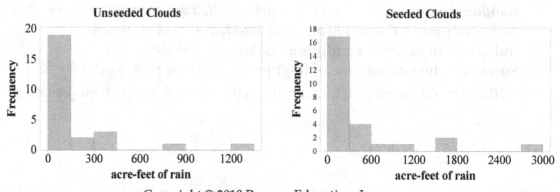

**b)** Re-expressing the cloud seeding data with logarithms makes each distribution much more unimodal and symmetric.

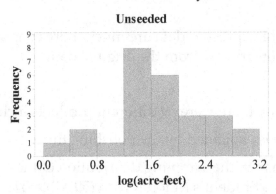

Unseeded

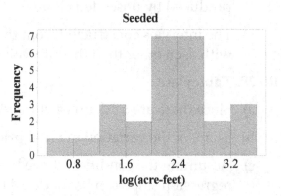

Seeded

**c)** Using logarithms to re-express the data allows the Nearly Normal Condition to be met. Since the conditions are satisfied, it is appropriate to model the sampling distribution of the difference in means with a Student's *t*-model, with 49.97 degrees of freedom (from the approximation formula). We will construct a two-sample *t*-interval, with 95% confidence.

$$\left(\overline{\log(y_s)} - \overline{\log(y_u)}\right) \pm t_{df}^* \sqrt{\frac{\left(SD\left(\log(y_s)\right)\right)^2}{n_s} + \frac{\left(SD\left(\log(y_u)\right)\right)^2}{n_u}}$$

$$= (2.229749 - 1.7330111) \pm t_{49.97}^* \sqrt{\frac{0.69465993^2}{26} + \frac{0.7130453^2}{26}}$$

$$\approx (0.10459974,\ 0.88887595)$$

We should certainly let technology do our calculating for us. Since we are using common logs, this is an interval of exponents of 10. We are carrying a large number of decimal places in our calculations to make the formula provide the correct outcomes, but a computer handles this easily.

Lower limit: $10^{0.10459974} \approx 1.27$ Upper Limit: $10^{0.88887595} \approx 7.74$

We are 95% confident the mean amount of rainfall produced by seeded clouds is between 1.27 acre-feet and 7.74 acre-feet more than the mean amount of rainfall produced by unseeded clouds.

As usual, answers will vary for the bootstrap interval. Based on the re-expressed data, our bootstrap interval for the difference in the logarithm of rainfall amount is (0.14346916, 0.87844334).

Lower limit: $10^{0.14346916} \approx 1.39$ Upper Limit: $10^{0.87844334} \approx 7.56$

We are 95% confident the mean amount of rainfall produced by seeded clouds is between 1.39 acre-feet and 7.56 acre-feet more than the mean amount of rainfall produced by unseeded clouds.

The intervals constructed from the re-expressed data are much more consistent with each other than the intervals constructed from the original data.

## R5.29. Tableware.

a) Since there are 57 degrees of freedom, there were 59 different products included.

b) 84.5% of the variation in retail price is explained by the polishing time.

c) Assuming the conditions have been met, the sampling distribution of the regression slope can be modeled by a Student's $t$-model with $(59 - 2) = 57$ degrees of freedom. We will use a regression slope $t$-interval. For 95% confidence, use $t^*_{57} \approx 2.0025$, or estimate from the table $t^*_{50} \approx 2.009$.

$$b_1 \pm t^*_{n-2} \times SE(b_1) = 2.49244 \pm (2.0025) \times 0.1416 \approx (2.21, \ 2.78)$$

d) We are 95% confident that the average price increases between $2.21 and $2.78 for each additional minute of polishing time.

## R5.31. Wealth redistribution 2015.

$H_0$: Income level and feelings about wealth redistribution are independent.

$H_A$: There is an association between income level and feelings about wealth distribution.

**Counted data condition:** The data are counts.
**Randomization condition:** Although not specifically stated, the Gallup Poll was likely to be random.

| | Should Redistribute (Obs/Exp) | Should Not (Obs/Exp) | No Opinion (Obs/Exp) |
|---|---|---|---|
| High Income | 426 / 534.33 | 579 / 460.33 | 10 / 20.33 |
| Middle Income | 558 / 534.33 | 447 / 460.33 | 10 / 20.33 |
| Low Income | 619 / 534.33 | 355 / 460.33 | 41 / 20.33 |

**Expected cell frequency condition:** The expected counts are all greater than 5.

Under these conditions, the sampling distribution of the test statistic is $\chi^2$ on 4 degrees of freedom. We will use a chi-square test for independence.

$$\chi^2 = \sum_{all\,cells} \frac{(Obs - Exp)^2}{Exp} \approx 123.02, \text{ and the } P\text{-value} < 0.0001.$$

Since the *P*-value is low, we reject the null hypothesis. There is strong evidence of an association between income level and opinion on wealth redistribution. Examination of the components shows that the low-income respondents are more likely to approve of redistribution when compared to the high-income respondents.

**R5.33. Lefties and music.**

$H_0$: The proportion of right-handed people who can match the tone is the same as the proportion of left-handed people who can match the tone.

$$\left(p_L = p_R \ \text{ or } \ p_L - p_R = 0\right)$$

$H_A$ : The proportion of right-handed people who can match the tone is different from the proportion of left-handed people who can match the tone.

$$\left(p_L \neq p_R \ \text{ or } \ p_L - p_R \neq 0\right)$$

**Random condition:** Assume that the people tested are representative of all people.
**Independent samples condition:** The groups are not associated.
**Success/Failure condition:** $n\hat{p}$ (right) = 38, $n\hat{q}$ (right) = 38, $n\hat{p}$ (left) = 33, and $n\hat{q}$ (left) = 20 are all greater than 10, so the samples are both large enough.

Since the conditions have been satisfied, we will model the sampling distribution of the difference in proportion with a Normal model with mean 0 and standard deviation estimated by

$$SE_{\text{pooled}}\left(\hat{p}_L - \hat{p}_R\right) = \sqrt{\frac{\hat{p}_{\text{pooled}}\hat{q}_{\text{pooled}}}{n_L} + \frac{\hat{p}_{\text{pooled}}\hat{q}_{\text{pooled}}}{n_R}} = \sqrt{\frac{\left(\frac{71}{129}\right)\left(\frac{58}{129}\right)}{53} + \frac{\left(\frac{71}{129}\right)\left(\frac{58}{129}\right)}{76}} \approx 0.089 \ .$$

The observed difference between the proportions is:
0.6226 – 0.5 = 0.1226.

Since the *P*-value = 0.1683 is high, we fail to reject the null hypothesis. There is no evidence that the proportion of people able to match the tone differs between right-handed and left-handed people.

$$z = \frac{0.1226 - 0}{0.089}$$

$$z \approx 1.38$$

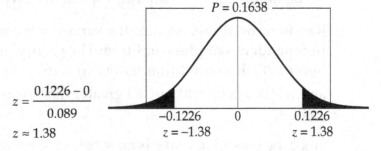

**R5.35. Twins.**

$H_0$: There is no association between duration of pregnancy and level of prenatal care.

$H_A$: There is an association between duration of pregnancy and level of prenatal care.

**Counted data condition:** The data are counts.

**Randomization condition:** Assume that these pregnancies are representative of all twin births.

**Expected cell frequency condition:** The expected counts are all greater than 5.

|  | Preterm (induced or Cesarean) (Obs / Exp) | Preterm (without procedures) (Obs / Exp) | Term or postterm (Obs / Exp) |
|---|---|---|---|
| Intensive | 18 / 16.676 | 15 / 15.579 | 28 / 28.745 |
| Adequate | 46 / 42.101 | 43 / 39.331 | 65 / 72.568 |
| Inadequate | 12 / 17.223 | 13 / 16.090 | 38 / 29.687 |

Under these conditions, the sampling distribution of the test statistic is $\chi^2$ on 4 degrees of freedom. We will use a chi-square test for independence.

$$\chi^2 = \sum_{all\,cells} \frac{(Obs - Exp)^2}{Exp} \approx 6.14,$$

and the $P$-value $\approx 0.1887$.

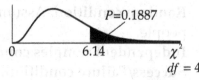

Since the $P$-value $\approx 0.1887$ is high, we fail to reject the null hypothesis. There is no evidence of an association between duration of pregnancy and level of prenatal care in twin births.

**R5.37. Retirement planning.**

$H_0$: The proportion of men who are "a lot behind schedule" in retirement planning is the same as the proportion of women. $\left(p_M = p_W \text{ or } p_M - p_W = 0\right)$

$H_A$: The proportion of men who are "a lot behind schedule" in retirement planning is lower than the proportion of women. $\left(p_M < p_W \text{ or } p_M - p_W < 0\right)$

**Random condition:** Assume the survey was conducted randomly.
**Independent samples condition:** The groups are not associated.
**Success/Failure condition:** $n\hat{p}$ (Men) = 267, $n\hat{q}$ (Men) = 455, $n\hat{p}$ (Women) = 301, and $n\hat{q}$ (Women) = 400 are all greater than 10, so the samples are both large enough.

Since the conditions have been satisfied, we will model the sampling distribution of the difference in proportion with a Normal model with mean 0 and standard deviation estimated by

$$SE_{pooled}\left(\hat{p}_M - \hat{p}_W\right) = \sqrt{\frac{\hat{p}_{pooled}\hat{q}_{pooled}}{n_M} + \frac{\hat{p}_{pooled}\hat{q}_{pooled}}{n_W}} = \sqrt{\frac{\left(\frac{568}{1423}\right)\left(\frac{855}{1423}\right)}{722} + \frac{\left(\frac{568}{1423}\right)\left(\frac{855}{1423}\right)}{701}} \approx 0.0260.$$

(We could also choose to use the unpooled standard error.)

The observed difference between the proportions is: 0.3698 – 0.4294 = 0.0596.

Since the *P*-value = 0.0109 is low, we reject the null hypothesis. There is evidence that the proportion of women who will say they are "a lot behind schedule" for retirement planning is higher than the percentage of men that would say the same thing.

$$z = \frac{(\hat{p}_M - \hat{p}_W) - 0}{SE(\hat{p}_M - \hat{p}_W)}$$

$$z = \frac{-0.0596}{0.0260} \approx -2.29$$

**R5.39. Eye and hair color.**

a) This is an attempt at linear regression. Regression inference is meaningless here, since eye and hair color are categorical variables.

b) This is an analysis based upon a chi-square test for independence.

H₀: Eye color and hair color are independent.

Hₐ: There is an association between eye color and hair color.

Since we have two categorical variables, this analysis seems appropriate. However, if you check the expected counts, you will find that 4 of them are less than 5. We would have to combine several cells in order to perform the analysis. (Always check the conditions!)

Since the value of chi-square is so high, it is likely that we would find an association between eye and hair color, even after the cells were combined. There are many cells of interest, but some of the most striking differences that would not be affected by cell combination involve people with fair hair. Blonds are likely to have blue eyes, and not likely to have brown eyes. Those with red hair are not likely to have brown eyes. Additionally, those with black hair are much more likely to have brown eyes than blue.

**R5.41. Cereals and fiber.**

a) For the simple regression: The model predicts that calories increase by about 1.26 calories per gram of carbohydrate.

For the multiple regression: After allowing for the effects of *fiber* in these cereals, the model predicts that calories increase at the rate of about 0.83 calories per gram of carbohydrate.

b) After accounting for the effect of *fiber*, the effect of *carbo* is no longer statistically significant.

c) I would like to see a scatterplot showing the relationship of *carbo* and *fiber*. I suspect that they may be highly correlated, which would explain why after accounting for the effects of *fiber*, there is little left for *carbo* to model. It would be good to know the $R^2$ value between the two variables.

**R5.43. Cereals with bran.**

These three cereals are highly influential in the relationship between *carbo* and *fiber*, are creating a strong, negative collinearity between them. Without these points, there is little association between these variables.

**R5.45. Togetherness.**

a) $H_0$: There is no linear relationship number of meals eaten as a family and grades. $(\beta_1 = 0)$

$H_A$: There is a linear relationship. $(\beta_1 \neq 0)$

Since the conditions for inference are satisfied (given), the sampling distribution of the regression slope can be modeled by a Student's *t*-model with (142 – 2) = 140 degrees of freedom. We will use a regression slope *t*-test. The equation of the line of best fit for these data points is: $\widehat{GPA} = 2.7288 + 0.1093(Meals / Week)$.

$$t = \frac{b_1 - \beta_1}{SE(b_1)}$$

$$t = \frac{0.1093 - 0}{0.0263}$$

$$t \approx 4.16$$

The value of $t \approx 4.16$. The *P*-value of less than 0.0001 means that the association we see in the data is unlikely to occur by chance. We reject the null hypothesis, and conclude that there is strong evidence of a linear relationship between grades and the number of meals eaten as a family. Students whose families eat together relatively frequently tend to have higher grades than those whose families don't eat together as frequently.

b) This relationship would not be particularly useful for predicting a student's grade point average. $R^2 = 11.0\%$, which means that only 11% of the variation in GPA can be explained by the number of meals eaten together per week.

c) These conclusions are not contradictory. There is strong evidence that the slope is not zero, and that means strong evidence of a linear relationship. This does not mean that the relationship itself is strong, or useful for predictions.

**R5.47. Juvenile offenders.**

a) **Randomization condition:** We will assume that the youths studied are representative of other youths that might receive this therapy.
**10% condition:** 125 and 125 are less than 10% of all such youths.
**Independent samples condition:** The groups are independent.
**Success/Failure condition:** $n\hat{p}$ (Ind) = 19, $n\hat{q}$ (Ind) = 106, $n\hat{p}$ (MST) = 5, and $n\hat{q}$ (MST) = 120 are all not greater than 10, but with only one equal to 5, the test two-proportion interval should be reliable.

Since the conditions have been satisfied, we will find a two-proportion z-interval.

$$\left(\hat{p}_I - \hat{p}_M\right) \pm z^* \sqrt{\frac{\hat{p}_I \hat{q}_I}{n_I} + \frac{\hat{p}_M \hat{q}_M}{n_M}}$$

$$= \left(\tfrac{19}{125} - \tfrac{5}{125}\right) \pm 2.576 \sqrt{\frac{\left(\tfrac{19}{125}\right)\left(\tfrac{106}{125}\right)}{125} + \frac{\left(\tfrac{5}{125}\right)\left(\tfrac{120}{125}\right)}{125}} = (0.0178, 0.206)$$

We are 99% confident that the percentage of violent felony arrest among juveniles who receive individual therapy is between 1.78 and 20.6 percentage points higher than the percentage of violent felony arrest among juveniles who receive MST.

b) Since the entire interval is above 0, we can conclude that MST is successful in reducing the proportion of juvenile offenders who commit violent felonies. The population of interest is adolescents with mental health problems.

**R5.49. Diet.**

$H_0$: Cracker type and bloating are independent.

$H_A$: There is an association between cracker type and bloating.

**Counted data condition:** The data are counts.
**Randomization condition:** Assume that these women are representative of all women.
**Expected cell frequency condition:** The expected counts are all (almost!) greater than 5.

Under these conditions, the sampling distribution of the test

| | Bloat | |
|---|---|---|
| | Little/None (Obs / Exp) | Moderate/Severe (Obs / Exp) |
| Bran | 11 / 7.6471 | 2 / 5.3529 |
| Gum Fiber | 4 / 7.6471 | 9 / 5.3529 |
| Combination | 7 / 7.6471 | 6 / 5.3529 |
| Control | 8 / 7.0588 | 4 / 4.9412 |

statistic is $\chi^2$ on 3 degrees of freedom. We will use a chi-square test for independence.

$$\chi^2 = \sum_{all\,cells} \frac{(Obs - Exp)^2}{Exp} \approx 8.23, \text{ and the } P\text{-value} \approx 0.0414.$$

$P = 0.0414$

$0 \qquad 8.23 \quad \chi^2$

$df = 3$

Since the P-value is low, we reject the null hypothesis. There is evidence of an association between cracker type and bloating. The gum fiber crackers had a higher rate of moderate/severe bloating than expected. The company should head back to research and development and address the problem before attempting to market the crackers.

**R5.51. Hearing.**

**Paired data assumption:** The data are paired by subject.
**Randomization condition:** The order of the tapes was randomized.
**Normal population assumption:** The histogram of differences between List A and List B is roughly unimodal and symmetric.

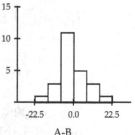

Since the conditions are satisfied, the sampling distribution of the difference can be modeled with a Student's *t*-model with 24 – 1 = 23 degrees of freedom. We will find a paired *t*-interval, with 95% confidence.

$$\bar{d} \pm t^*_{n-1}\left(\frac{s_d}{\sqrt{n}}\right) = -0.\overline{3} \pm t^*_{23}\left(\frac{8.12225}{\sqrt{24}}\right) \approx (-3.76, 3.10)$$

We are 95% confident that the mean difference in the number of words a person might misunderstand using these two lists is between –3.76 and 3.10 words. Since 0 is contained in the interval, there is no evidence to suggest that that the two lists are different for the purposes of the hearing test when there is background noise. It is reasonable to think that the two lists are still equivalent.

# Parts I-V Cumulative Review Exercises

1. **Igf.**

   a) The distribution of insulin-like growth factor is bimodal and skewed to the right, with modes at approximately 200 and 400 $\mu g / l$. The median is 316 $\mu g / l$, with the middle 50% of observations between 203.5 and 464 $\mu g / l$. The lowest observation was 25 $\mu g / l$ and the highest observation was 915 $\mu g / l$.

   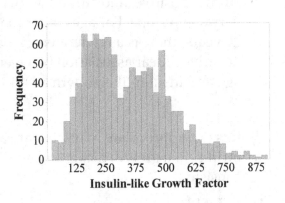

   b) For both males and females, the distributions of insulin-like growth factor are very similar to the overall distribution. Both are bimodal and skewed to the right, with modes at approximately 200 and 400 $\mu g / l$. The median for males is 280 $\mu g / l$, while the median for females is higher, at 352 $\mu g / l$. The middle 50% of observations for males is between 176 and 430.75 $\mu g / l$. For females, the middle 50% of the observations is generally higher, between 232.50 and 484 $\mu g / l$. The extreme observations for males and females are approximately equal with minimum values of 29 and 25 $\mu g / l$ and maximum values of was 915 and 914 $\mu g / l$, for males and females, respectively. Both groups have high outliers.

   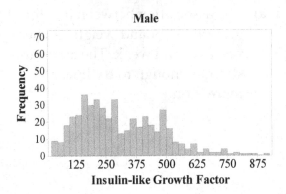

   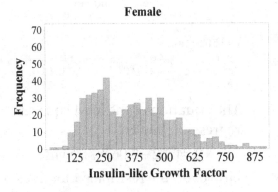

   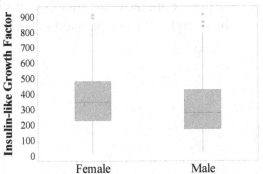

**c)** For ages younger than 20 years, the association between age and insulin-like growth factor is moderately strong, positive, and curved, with a very steep slope. For ages older than 20 years, there is a moderately strong, negative, linear association between age and insulin-like growth factor.

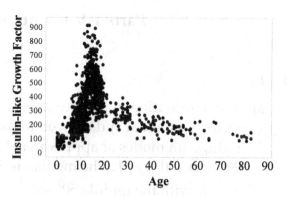

**d)** It is not appropriate to use linear regression to model the association between insulin-like growth factor and age. The relationship is neither linear nor consistent.

**3. More Igf13.**

**a)** The association between insulin-like growth factor and weight is linear, positive, and weak. The association is straight enough to try linear regression.

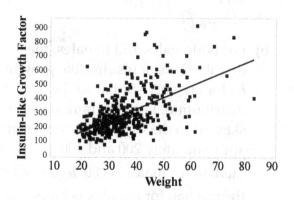

```
Model Summary
      S     R-sq   R-sq(adj)
 121.264   31.17%    31.00%

Coefficients
Term       Coef      SE Coef   T-Value  P-Value
Constant   2.94202   21.81     0.135    0.8928
weight     8.06787   0.5972    13.5     <0.0001
```

The equation of the least squares regression line is

$\widehat{Igf} = 2.94202 + 8.06787\,weight$.

However, the residuals plot does not show a constant variance. We will have to try a re-expressed model.

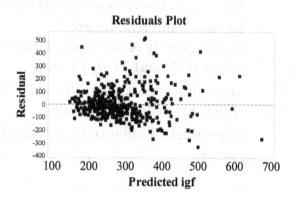

Re-expressing the insulin-like growth factor with logarithms improves the model.

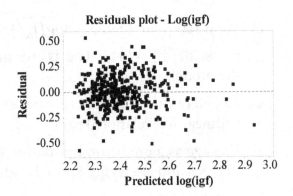

Residuals plot - Log(igf)

```
Model Summary
      S    R-sq   R-sq(adj)
 0.170465  29.83%    29.66%


Coefficients
Term         Coef      SE Coef   T-Value    P-Value
Constant    2.02218    0.0307     65.95     <0.0001
weight      0.0109898  0.000840   13.09     <0.0001
```

The equation of the least squares regression line is

$$\widehat{\log(Igf)} = 2.02218 + 0.0109898\,weight.$$

**b)** To add the variable *sex*, we must code it. Here we used male = 1, female = 0.

```
Model Summary
      S    R-sq   R-sq(adj)
 0.164999  34.42%    34.10%


Coefficients
Term          Coef      SE Coef   T-Value    P-Value
Constant    2.05963     0.0305     67.51     <0.0001
weight      0.011009    0.000813   13.55     <0.0001
sex        -0.087745    0.0165     -5.30     <0.0001
```

The interpretation of the coefficient of the logarithmic re-expression is difficult to interpret meaningfully. If we go back the linear model, and include *sex* as a predictor, the interpretation of the coefficient is that boys have, on average, an igf level that is 61.6 units lower than girls after allowing for the effects of weight.

A plot of the residuals shows that the slopes for boys and girls are not parallel for regression on log(igf). So, a careful student might conclude that this isn't an appropriate regression.

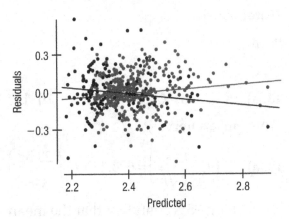

   c) For a 100-pound female, $\widehat{\log(Igf)} = 2.05963 + 0.011(100) \approx 3.15963$. In natural units, $10^{3.15963} \approx 1444$. This is an extrapolation well beyond the data used to build the model.

   d) No, all of the coefficients are not statistically significant. Height and weight are collinear (correlation 0.887).

   e) Use age as a predictor. It is odd to leave that out when we have ages from infants to teens and are trying to use height and weight as predictors.

5. **Hospital variables.**

   a) Two-sided one-proportion $z$-test for a proportion, $H_0 : z = 0.50$.

   b) Two-sided two-proportion $z$-test for difference between proportions.

   c) One-sided two-sample $t$-test for difference between means.

   d) Two-sided two-sample $t$-test for difference between means.

   e) Two-sided linear regression $t$-test for slope of the line of best fit predicting 2018 appointments from 2017 appointments.

7. **Life expectancy and literacy.**

   a) The scatterplot looks straight enough to use the linear regression model, $\widehat{LifeExp} = 72.3949 - 1.2960Illiteracy$. When the illiteracy rate is 0.70, the life expectancy is predicted to be: $\widehat{LifeExp} = 72.3949 - 1.2960(0.70) \approx 71.49$ years.

   b) IV

   c) $r = \sqrt{R^2} = \sqrt{0.3462} \approx 0.588$

   d) IV

   e) I

**Hotel maids.**

9. a

11. c

13. a) A    b) C    c) E    d) D    e) B

**Olympic archery.**

15. a) $\bar{y} \pm t^*_{n-1}\left(\dfrac{s}{\sqrt{n}}\right) = 624.41 \pm t^*_{63}\left(\dfrac{28.52}{\sqrt{64}}\right) \approx (617.28,\ 631.53)$.

   I am 95% confident that the mean women's archery seeding score is between 617.28 and 631.53.

**b)** This question does not relate to the confidence interval constructed in part a. Confidence intervals are about means, not individual performances. With a mean score of 624.41 and a standard deviation of 28.52, a score of 660 is only about 1.25 standard deviations above the mean. Zhang Juanjuan's score is high, but not extraordinarily high.

**17.** $H_0$: The mean seeding score is the same for females and males.

$$\left(\mu_F = \mu_M \ \text{ or } \ \mu_F - \mu_M = 0\right)$$

$H_A$ The mean seeding score is different for females and males.

$$\left(\mu_F \neq \mu_M \ \text{ or } \ \mu_F - \mu_M \neq 0\right)$$

**Independent groups assumption:** Female scores should not affect male scores.

**Randomization condition:** We will assume that these archery scores are representative of the scores of female and male Olympic archers.

**Nearly Normal condition:** The distributions of seeding scores are both skewed to the right, but with sample sizes of 64 for both groups, the Central Limit Theorem will allow us to continue.

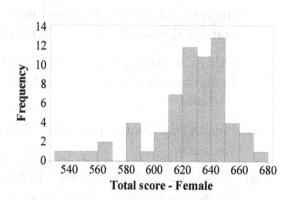

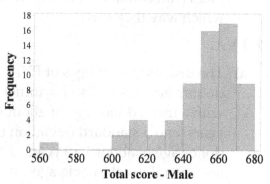

Since the conditions are satisfied, it is appropriate to model the sampling distribution of the difference in means with a Student's $t$-model, with 115.77 degrees of freedom (from the approximation formula). We will perform a two-sample $t$-test. We know:

$$\bar{y}_F = 624.40625 \qquad \bar{y}_M = 651.09375$$
$$s_F = 28.521626 \qquad s_M = 20.991093$$
$$n_F = 64 \qquad n_M = 64$$

The sampling distribution model has mean 0, with standard error:

$$SE(\bar{y}_F - \bar{y}_M) = \sqrt{\frac{28.521626^2}{64} + \frac{20.991093^2}{64}} \approx 4.4266754 \ .$$

The observed difference between the mean scores is
$624.40625 - 651.09375 \approx -26.6875$

Since the *P*-value < 0.0001 is very low, we reject the null hypothesis. There is strong evidence that the mean seeding scores for males and females are different. These data suggest that males have significantly higher mean seeding scores.

$$t = \frac{(\bar{y}_F - \bar{y}_M) - (0)}{SE(\bar{y}_F - \bar{y}_M)}$$

$$t \approx \frac{-26.6875}{4.426674}$$

$$t \approx -6.03$$

### 19. Belmont stakes.

**a)** We are 95% confident that the mean clockwise speed is between 1.76 and 2.54 miles per hour slower than the mean counterclockwise speed. Since 0 is not in the interval, these data provide evidence that the clockwise speeds are statistically significantly slower, on average.

**b)** Oddly, horses appear to run faster in longer races, with the 1.625-mile races being an exception. But that doesn't make sense, so there may be a lurking variable.

**c)** I expect the confidence interval to be narrower. There is a bigger effect and less variation.

**d)** Year is the lurking variable. In fact, horses have gotten faster, so more recent races (run counterclockwise) were faster. There is no evidence that horses care which way they run.

### 21. PVA.

**a)** The distribution of ages of PVA donors is unimodal and symmetric, with a mean donor age of about 61.5 years, and a standard deviation on donor age of about 15.0 years. There are several questionable ages. It seems unlikely that there were donors of ages 4, 13, and 15 years. Perhaps someone donated to the organization on their behalf.

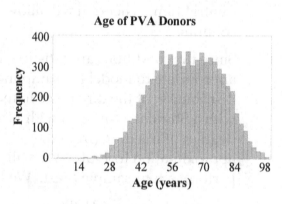

We have taken a random sample of the entire donor database, and the distribution of ages is roughly unimodal and symmetric, so we can use a one-sample *t*-interval to estimate the mean age of all donors to PVA.

$$\bar{y} \pm t^*_{n-1}\left(\frac{s}{\sqrt{n}}\right) = 61.524 \pm t^*_{7599}\left(\frac{15.030}{\sqrt{7600}}\right) \approx (61.186, \ 61.862)$$

We are 95% confident that the mean age of all PVA donors is between 61.186 and 61.862 years.

**b)** H₀: The mean age of homeowners among PVA donors is the same as the mean age of non-homeowners among PVA donors. $\left(\mu_H = \mu_N \text{ or } \mu_H - \mu_N = 0\right)$

H_A The mean age of homeowners among PVA donors higher than the mean age of non-homeowners among PVA donors. $\left(\mu_H > \mu_N \text{ or } \mu_H - \mu_N > 0\right)$

**Independent groups assumption:** Homeowners ages should not affect non-homeowners ages.
**Randomization condition:** This data set is a random sample of all donors in the PVA database.
**Nearly Normal condition:** The distributions of ages for both groups are roughly unimodal and symmetric.

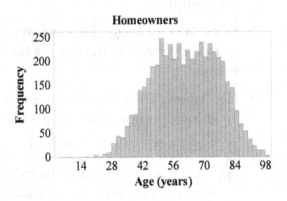

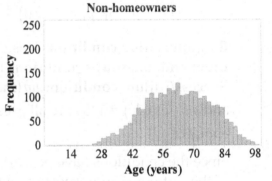

Since the conditions are satisfied, it is appropriate to model the sampling distribution of the difference in means with a Student's *t*-model, with 4896 degrees of freedom (from the approximation formula). We will perform a two-sample *t*-test. We know:

$$\bar{y}_H = 61.234 \qquad \bar{y}_N = 62.121$$
$$s_H = 615.016 \qquad s_N = 15.044$$
$$n_H = 5120 \qquad n_N = 2480$$

The sampling distribution model has mean 0, with standard error:

$$SE(\bar{y}_H - \bar{y}_N) = \sqrt{\frac{15.016^2}{5120} + \frac{15.044^2}{2480}} \approx 0.368 .$$

The observed difference between the mean scores is
61.234 – 62.121 ≈ –0.887

Since the *P*-value = 0.992 is high, we fail to reject the null hypothesis. There is no evidence that the mean age of homeowners is higher than the mean age of non-homeowners among PVA donors. (Did you make it this far without realizing that the mean age of homeowners is actually *lower* in this sample?)

$$t = \frac{(\bar{y}_F - \bar{y}_M) - (0)}{SE(\bar{y}_F - \bar{y}_M)}$$

$$t \approx \frac{-0.887}{0.368}$$

$$t \approx -2.41$$

Since the mean age of donors overall is quite high, this discrepancy could be caused by many of the donors having retired. These retirees may no longer own homes.

c) $H_0$ : Among PVA donors, the proportion of males who own homes is the same as the proportion of females who own homes.

$$\left(p_M = p_F \ \text{ or } \ p_M - p_F = 0\right)$$

$H_A$ : Among PVA donors, the proportion of males who own homes is greater than the proportion of females who own homes.

$$\left(p_M > p_F \ \text{ or } \ p_M - p_F > 0\right)$$

**Randomization condition:** This data set is a random sample of PVA donors.
**Independent groups condition:** The groups are not associated.
**Success/Failure condition:** $n\hat{p}$ (male) =2287, $n\hat{q}$ (male) = 1026, $n\hat{p}$ (female) =2833, and $n\hat{q}$ (female) = 1454 are all greater than 10, so the samples are both large enough.

Since the conditions have been satisfied, we will model the sampling distribution of the difference in proportion with a Normal model with mean 0 and standard deviation estimated by

$$SE(\hat{p}_{None} - \hat{p}_{Dep}) = \sqrt{\frac{\hat{p}_M \hat{q}_M}{n_M} + \frac{\hat{p}_F \hat{q}_F}{n_F}} = \sqrt{\frac{\left(\frac{2287}{3313}\right)\left(\frac{1026}{3313}\right)}{3313} + \frac{\left(\frac{2833}{4287}\right)\left(\frac{1454}{4287}\right)}{4287}} \approx 0.010808 \, .$$

(We could have used the pooled standard error as well.)

The observed difference between the proportions is:
0.690311 − 0.660835 = 0.0294758.

$$z = \frac{\hat{p}_M - \hat{p}_F}{SE(\hat{p}_M - \hat{p}_F)}$$

Since the *P*-value = 0.006 is low, we reject the null hypothesis. There is evidence to suggest that, among PVA donors, the proportion of males who own homes is higher than the proportion of females who own homes.

$$z = \frac{0.0294758}{0.010808}$$

$$z \approx 2.73$$

We could have used a $\chi^2$ test for independence as well. Recall that the test for independence is two-sided by nature, so we would have to test a hypothesis of a difference in homeownership rates, rather than the hypothesis that males have a higher homeownership rate. Nevertheless, our conclusions would have been the same.

**d)** The distribution of current gift of PVA donors is unimodal and extremely skewed to the right. The median gift is $8, but gifts are highly variable. The lowest current gift is $0 (with over a thousand current donors giving nothing this year), and the highest current gift is $200. There are many high outliers.

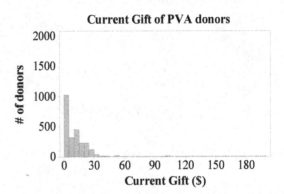

**Current Gift of PVA donors**

# Normal Model

- Large Populations
- $p$ can't be close to $0$ or $1$

- $p = $ mean $\quad \sqrt{\dfrac{pq}{n}} = $ sd

- Don't sample $+ 10\%$ of population
- $10$ successes $+$ failures

$$SD(\hat{p}) = \sqrt{\dfrac{pq}{n}}$$

$$SE(\hat{p}) = \sqrt{\dfrac{\hat{p}\hat{q}}{n}}$$